T0210704

# SpringerBriefs in Computer Science

SpringerBriefs present concise summaries of cutting-edge research and practical applications across a wide spectrum of fields. Featuring compact volumes of 50 to 125 pages, the series covers a range of content from professional to academic.

Typical topics might include:

- A timely report of state-of-the art analytical techniques
- A bridge between new research results, as published in journal articles, and a contextual literature review
- A snapshot of a hot or emerging topic
- An in-depth case study or clinical example
- A presentation of core concepts that students must understand in order to make independent contributions

Briefs allow authors to present their ideas and readers to absorb them with minimal time investment. Briefs will be published as part of Springer's eBook collection, with millions of users worldwide. In addition, Briefs will be available for individual print and electronic purchase. Briefs are characterized by fast, global electronic dissemination, standard publishing contracts, easy-to-use manuscript preparation and formatting guidelines, and expedited production schedules. We aim for publication 8–12 weeks after acceptance. Both solicited and unsolicited manuscripts are considered for publication in this series.

**Indexing: This series is indexed in Scopus, Ei-Compendex, and zbMATH **

Guangtao Xue • Yi-Chao Chen • Feng Lyu •
Minglu Li

# Robust Network
# Compressive Sensing

 Springer

Guangtao Xue
Shanghai, China

Yi-Chao Chen
Shanghai, China

Feng Lyu
Hunan, China

Minglu Li
Jinhua, China

ISSN 2191-5768          ISSN 2191-5776   (electronic)
SpringerBriefs in Computer Science
ISBN 978-3-031-16828-4          ISBN 978-3-031-16829-1    (eBook)
https://doi.org/10.1007/978-3-031-16829-1

This Springer imprint is published by the registered company Springer Nature Switzerland AG
The registered company address is: Gewerbestrasse 11, 6330 Cham, Switzerland

# Preface

The networks are constantly generating a wealth of rich and diverse information. This information creates exciting opportunities for network analysis and provides insight into the complex interactions between network entities. However, network analysis often faces the problems of (1) under-constrained, where there is too little data due to feasibility and cost issues in collecting data, or (2) over-constrained, where there is too much data, so the analysis becomes unscalable. Compressive sensing is an effective technique to solve both problems. It utilizes the underlying data structure for analysis. Specifically, to solve the under-constrained problem, compressive sensing technologies can be applied to reconstruct the missing elements or predict the future data; to solve the over-constraint problem, compressive sensing technologies can be applied to identify significant elements.

To well support compressive sensing in network data analysis, a robust and general framework is needed to support diverse applications, yet this is challenging for real-world data where noise, anomalies, and lack of synchronization are common. First, the number of unknowns for network analysis can be much larger than the number of measurements we have. For example, traffic engineering requires knowing the complete traffic matrix between all source and destination pairs in order to properly configure traffic and avoid congestion. However, measuring the flow between all source and destination pairs is very expensive or even infeasible. Reconstructing data from a small number of measurements is an under-constrained problem. In addition, real-world data are complex and heterogeneous and often violate the low-level assumptions required by existing compressive sensing techniques. Such violations significantly reduce the applicability and effectiveness of existing compressive sensing methods. Third, synchronization of network data reduces the data ranks and increases spatial locality. However, periodic time series exhibit not only misalignment but also different frequencies, which makes it difficult to synchronize data in the time and frequency domains.

In this monograph, we investigate compressive sensing techniques to provide a robust and general framework for network data analytics. Our goal is to build a compressive sensing framework for missing data interpolation, anomaly detection, data segmentation, and activity recognition, and to demonstrate its benefits. In

Chap. 1, we introduce compressive sensing, including its definition, limitation, and how it supports different network analysis applications. In Chap. 2, to demonstrate the feasibility of compressive sensing in network analytics, we apply it to detect anomalies in the customer care call dataset from a Tier 1 ISP in the United States. A regression-based model is applied to find the relationship between calls and events. We show that compressive sensing is effective in identifying important factors and can leverage the low-rank structure and temporal stability to improve the detection accuracy. In Chap. 3, we show that there are several challenges in applying compressive sensing to real-world data. Understanding the reasons behind the challenges is important for designing methods and mitigating their impact. Therefore, we analyze a wide range of real-world traces, and our analysis shows that there are different factors that contribute to the violation of the low-rank property in real data. In particular, we find that (1) noise, errors, and anomalies, and (2) asynchrony in the time and frequency domains lead to network-induced ambiguity and can easily cause low-rank matrices to become higher-ranked. To address the problem of noise, errors, and anomalies, in Chap. 4, we propose a robust compressive sensing technique. It explicitly accounts for anomalies by decomposing real-world data represented in matrix form into a low-rank matrix, a sparse anomaly matrix, an error term, and a small noise matrix. In Chap. 5, to address the problem of lack of synchronization, we propose a data-driven synchronization algorithm. It can eliminate misalignment while taking into account the heterogeneity of real-world data in both time and frequency domains. The data-driven synchronization can be applied to any compressive sensing technique and is general to any real-world data. We show that the combination of the two techniques can reduce the ranks of real-world data, improve the effectiveness of compressive sensing, and have a wide range of applications.

Shanghai, China                                                             Guangtao Xue
Shanghai, China                                                              Yi-Chao Chen
Hunan, China                                                                      Feng Lyu
Zhejiang, China                                                                   Minglu Li

# Contents

# Acronyms

| | |
|---|---|
| ADL | Activities for daily living |
| ADM | Alternating direction method |
| AP | Access point |
| ColRandLoss | Drop random columns |
| CSI | Channel state information |
| DDoS | Distributed denial of service |
| ETT | Expected transmission time |
| EWMA | Exponential weighted moving average |
| FLP | Florida/Puerto Rico area |
| IVR | Interactive voice response |
| KNN | K-nearest neighbors |
| LENS | The algorithm to decompose a matrix into a low-rank matrix, a sparse anomaly matrix, an error matrix, and a dense but small noise matrix |
| MRC | Maximum ratio combining |
| NCCO | National call center operations |
| NMAE | Normalized mean absolute error |
| PCA | Principal component analysis |
| PureRandLoss | Elements in a matrix are dropped independently with a random loss rate |
| RowRandLoss | Drop random rows |
| RSSI | Received signal strength indicator |
| RT | Retweets |
| SRMF | Sparsity regularized matrix factorization |
| STFT | Short-time Fourier transform |
| SVD | Singular value decomposition |
| TF-IDF | Term frequency-inverse document frequency |
| xxElemRandLoss | xx% of rows in a matrix are selected and the elements in these selected rows are dropped with a probability |

xxElemSyncLoss    Intersection of xx% of rows and p% of columns in a matrix are
                  dropped
xxTimeRandLoss    xx% of columns in a matrix are selected and the elements in
                  these selected columns are dropped with a probability

# Chapter 1
# Introduction

Wireless networks and sensor networks are constantly generating an enormous amount of rich and diverse information. Such information creates exciting opportunities for network analytics. Network analytics can provide deep insights into the complex interactions among network entities, and has a wide range of applications in wireless networks across all protocol layers. Example applications include spectrum sensing, channel estimation, channel feedback compression, multi-access channel design, data aggregation, network coding, wireless video coding, anomaly detection, activity recognition, and localization.

There are many challenges to enable effective network analytics:

- **Under-constraint**: The number of unknown factors for network analytics can be much larger than the number of measurements we made. For example, traffic engineering requires knowing the complete traffic matrix between all source and destination pairs in order to properly provision traffic and avoid congestion. However, it's very expensive, if not infeasible, to measures the traffic between all source and destination pairs. Reconstructing data from few measurements is an under-constraint problem.
- **Over-constraint**: Even we are fortunate to have enough measurements, network analytics can still be challenging because it may try to describe the biased measurements or random noise instead of the underlying relationship. Moreover, the large amount of data may make the analytics unscalable.

Compressive sensing is an effective technique to solve the problems. It has recently attracted considerable attention in statistics, approximation theory, information theory, and signal processing. It leverages the presence of structure and redundancy in real data for interpolation and analysis. In particular, several effective heuristics have been proposed to explicitly exploit the sparse or low-rank nature of empirically obtained matrices (i.e., a matrix can be approximated as a product of two factor matrices with few columns). Meanwhile, the mathematical theory of compressive sensing has also advanced to the point where the optimality of many of these heuristics has been proven under certain technical conditions on the matrices of interest.

There are several advantages for applying compressive sensing to network analytics.

- Recover missing data: High cost in collecting complete data, failures in measurement systems, and network losses can lead to missing data. On the other hand, many network tasks require complete data to operate properly. Compressive sensing can be applied to interpolate missing data where there is too little data for analytics.
- Find the underlying data structure: Compressive sensing reconstruct data by utilizing the underlying data structure. By designing the penalty terms carefully, we can find the intrinsic trend in the data and avoid over-fitting. Moreover, we can also remove noise and detect anomalies in the data because the residual which is inconsistent with the underlying data structure can be identified.
- Scalability and power efficiency: Compressive sensing requires few measurements to reconstruct the data so reduces the cost for measurement and data collection. This is important for network analytics to scale in large networks and save power in sensor or mobile networks.

## 1.1   Apply Compressive Sensing to Real-World Data

We demonstrate the feasibility of compressive sensing in network analytics by applying it to detect events in a customer care calls dataset collected by a tier-1 ISP in US.

Customer care calls serve as a direct channel for a service provider to learn feedbacks from their customers. They reveal details about the nature and impact of major events and problems observed by customers. By analyzing the customer care calls, a service provider can detect important events to speed up problem resolution. However, automating event detection based on customer care calls poses several significant challenges. First, the relationship between customers' calls and network events is blurred because customers respond to an event in different ways. Second, customer care calls can be labeled inconsistently across agents and across call centers, and a given event naturally give rise to calls spanning a number of categories. Third, many important events cannot be detected by looking at calls in one category. How to aggregate calls from different categories for event detection is important but challenging. Lastly, customer care call records have high dimensions (e.g., thousands of categories in our dataset).

We propose a systematic method for detecting events in a major cellular network using customer care call data. It consists of three main components: (i) using a regression approach that exploits temporal stability and low-rank properties to automatically learn the relationship between customer calls and major events, (ii) reducing the number of unknowns by clustering call categories and using L1 norm minimization to identify important categories, and (iii) employing multiple classifiers to enhance the robustness against noise and different response time. For the detected events, we leverage Twitter social media to summarize them and to locate the impacted regions. We show that compressive sensing is effective in identifying important factors and can leverage the low-rank structure and temporal stability of the data to improve the detection accuracy.

## 1.2   Understand the Limitation

While applying compressive sensing to the real-world data, we identify several challenges. One of the challenges is that real-world data are complicated and heterogeneous, and often violate the low-rank assumption required by existing compressive sensing techniques. Such violation significantly reduces the applicability and effectiveness of existing compressive sensing approaches. It is important to understand reasons behind the violation to design methods and mitigate the impact. Therefore, we analyze a wide range of real-world traces and our analysis reveals that there are different factors that contribute to the violation of low-rank property in real data. In particular, we find the following factors lead to network-induced blurring, and can easily cause a low-rank matrix to become a much higher rank:

- **Noise, Errors, and Anomalies**: Noise, erroneous data, and anomalies are common in real-world network data. Anomalies and errors can hide non-anomaly-related data. It is challenging to distinguish genuine network structure and behavior of interest from anomalies and measurement imperfections in a robust and accurate fashion.
- **Lack of synchronization in time domain and frequency domain**: In a distributed network, different nodes have different clocks and may also span multiple time zones. The matrix formed by merging the measurements across different nodes is likely to have misalignment. Even with perfect clock synchronization, some network events take time to propagate and some network elements may see their effects earlier than the other elements. Therefore it is not sufficient to synchronize the clocks, but rather we should synchronize the underlying data. Moreover, in order to support a wide range of real-world data, we should take into account of both the time domain and frequency domain information arising from different sampling rates or speeds of movement.

## 1.3   Improve Compressive Sensing for Real-World Data

**Robust Compressive Sensing**   To address the problem due to the presence of measurement errors, noise, and anomalies, we develop *LENS decomposition*, a novel technique to accurately decompose network data represented in the form of a matrix into a Low-rank matrix, an Error matrix, a small Noise matrix, and a Sparse anomaly matrix. This decomposition naturally reflects the inherent structures of real-world data and is more general than existing compressive sensing techniques by removing the low-rank assumption and explicitly supporting anomalies. The problem can be generalized to incorporate domain knowledge, such as temporal stability, spatial locality, and/or initial estimation (e.g., obtained from a model). We formulate this problem as a convex optimization problem, and develop an Alternating Direction Method (ADM) to efficiently solve it.

Our approach has several nice properties: (i) it supports a wide range of matrices: with or without anomalies, and with or without low-rank structure, (ii) its parameters are either exactly computed or self tuned, (iii) it can incorporate domain knowledge, and (iv) it supports various network applications, such as missing value interpolation, prediction, and anomaly detection.

**Data-Driven Synchronization** To address the problem due to lack of synchronization in time and frequency domain, we propose a data-driven synchronization approach to explicitly remove misalignment while accounting for the time domain and frequency domain heterogeneity of the real-world data. We show synchronization improves interpolation accuracy, and allows us to exploit the correlation among different users' measurements to enhance segmentation and activity recognition.

We evaluate LENS and data-driven synchronization using a wide range of real-world data including network traces from traffic matrices in 3G, WiFi, and the Internet, channel state information (CSI) matrices, RSSI matrices in WiFi and sensor networks, and expected transmission time (ETT) traces from UCSB Meshnet, as well as activity traces from wearable devices. Our results show that (i) LENS significantly out-performs state-of-the-art compressive sensing methods, (ii) data-driven synchronization further reduces the matrix ranks and improves the interpolation performance of all compressive sensing methods, and (iii) data-driven synchronization enables group-based segmentation and activity recognition which significantly out-performs existing schemes which applied on an individual user at a time.

## 1.4  Aim of the Monograph

In summary, the aim of the monograph are as follows:

- We develop a systematic method to automatically detect anomalies in a cellular network using the customer care call data collected by a tier-1 ISP in US. Compressive sensing is used to enhance the scalability and improve the accuracy. Our approach out-performs the existing approach (i.e., a regular regression approach which only minimizes the fitting error) by 64%.
- We make important observations that real-world datasets often are not low-rank due to (i) the presence of measurement errors, noise, and anomalies, (ii) lack

of synchronization in time and frequency domain. To our knowledge, this is the *first* work that shows the impact of noise, anomalies, and synchronization in the matrix ranks, which has profound implication for compressive sensing.

- We propose a robust compressive sensing technique for real-world network datasets explicitly accounts for anomalies and measurement noise. The proposed technique out-performs the state-of-the-art approaches by 20.2–69.8% while applied in missing data interpolation, 17.7–34.6% in prediction, and 17.6% in anomaly detection.

- We develop a simple yet effective data-driven synchronization algorithm explicitly accounts for the time domain and frequency domanin heterogeneity in the real-world data. Our results show synchronizing the data to a common offset and frequency reduces the matrix ranks by up to 65%.

- We design algorithms that take advantage of synchronization for data interpolation, segmentation, and activity recognition. We reduces interpolation error by up to 79%, and significantly out-performs individual user based segmentation and activity recognition by 12–61% and 10–123%, respectively.

# Chapter 2
# Event Detection System

In this chapter, we show how compressive sensing technique can be applied to build an event detection system in a major cellular network using customer care call data.

## 2.1  Problem Formulation

**Background** This study uses data derived from operational records of calls to customer support centers of a major mobile service provider. We now outline how these records are generated at the service center, and then describe a subset of the records that are used for this study.

When a customer calls the customer support, the call will first reach an Interactive Voice Response (IVR) system, an automated system configured with pre-defined menu. Based on the selected menu, the customer's call is either self-served or routed to one of the customer care call centers to be answered by an agent. Work force management for customer call centers are often performed according to the type of plan the customer has (e.g., business or consumer, referred as "work group") and what type of issues the customer has (e.g., device, billing, performance issues).

Upon handling each customer care call, the call agent will open a case in the ticketing system, and label the case using a "three-level pre-defined categories" to indicate the customer's issue or need. Detailed notes are also input into the system based on the conversation with the customer. After the customer's need is satisfied and case is resolved, a "call resolution" code is also entered into the system. Although the detailed notes may provide more detailed information and help detect anomalies, they are used in this study due to the privacy issues and the challenge of using natural language to process them. Therefore, this chapter focuses on using pre-defined categories of the calls to detect anomalies.

© The Author(s), under exclusive license to Springer Nature Switzerland AG 2022
G. Xue et al., *Robust Network Compressive Sensing*, SpringerBriefs in Computer
Science, https://doi.org/10.1007/978-3-031-16829-1_2

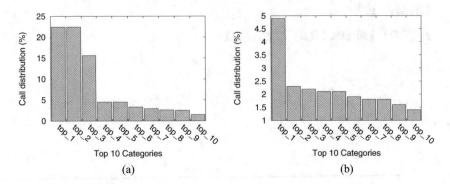

**Fig. 2.1** The normalized number of calls of *work group* and *call resolution*. (**a**) Work Group (141 categories). (**b**) Call Resolution (5394 categories)

**Customer Care Calls Dataset**  The data set for this study is derived from several million calls received at the service centers in 5 months during 2011. Each record in the dataset comprises the following subset of information related to a call: *work group*, *call resolution* (as described above) and the category ascribed to the call, comprising three levels customer need: *customer need level 1*, *customer need level 2*, and *customer need level 3*.

There are 141 work groups, 5394 call resolutions, 170 level-1 categories, 765 level-2 categories, and 2882 level-3 categories. Data do not contain any information concerning calls that did not progress beyond the IVR system. Figures 2.1 and 2.2 shows the normalized call distribution in the most popular 10 categories among *work group*, *call resolution* and 3 levels of *customer need*. We can observe that the top 10 categories account for 30–70% of calls. Moreover, top 10% categories account for 75–90% of calls. In spite of the significant amount of calls in top 10% categories, we cannot simply use these categories for anomaly detection because there are many anomalies dominated by the remaining 90% categories. For examples, "Technical" is a level-1 category which does not belong to the top 10% categories but it is one of the dominant factors in 73% of outage events observed from our dataset.

**Real Anomalies**  In addition, we get ground truth from National Call Center Operations (NCCO) reports, in which anomalies are marked manually by monitoring the patterns of customer calls, activity of the IVR system, and network traffic. This process is extremely time-consuming and vulnerable to human errors, which motivates us to develop an automatic anomaly detector.

Table 2.1 shows an example of NCCO report. Each anomaly in NCCO reports contains the following information: an *event status* indicates the anomaly is first reported (*initial*), *updated*, or *resolved*; a *business unit* and *primary system* indicate which aspect of system is impacted by the anomaly; and a *region* indicates the impacted region. A *start time* labels the time when the anomaly is observed, and *resolved time* labels the time when the support team reports that the anomaly has

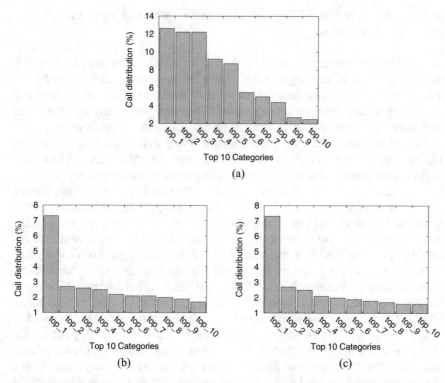

**Fig. 2.2** The normalized number of calls of three levels. (**a**) Level-1 (170 categories). (**b**) Level-2 (765 categories). (**c**) Level-3 (2882 categories)

**Table 2.1** An example of NCCO report

| |
|---|
| **Event Status**: Initial |
| **Business Unit(s)**: Mobility |
| **Primary System**: Mobility: GSM Voice Networks |
| **Region**: North East |
| **Start Time**: month day year - time |
| **Resolved Time**: Unknown |
| **Issue Description**: Boston customers may experience no service or degraded service in the coverage area of the cell sites affected. |

been resolved. In this example, the anomaly is just reported, so the *resolved time* is unknown. There is also a description of the anomaly, e.g., outage events or performance degradation. We designated the *start time* as the time at which an anomaly is detected. There could be a gap between the time that anomalies are observed in NCCO report and perceived by customers.

Since our approach uses only customer calls data to detect anomalies while the ground truth from NCCO reports is derived based on the more complete information (e.g., activity of the IVR system and network traffic), we do not expect our approach can detect all anomalies.

**Issues** The call records give us information about calls in different categories with different metrics. Each category/metric gives us one timeseries. Our goal is to automatically detect anomalies or events using all the timeseries. A natural approach is to detect sudden changes in one or more timeseries. But finding an appropriate aggregation of the timeseries for accurate anomaly detection is challenging.

Simply aggregating all of them does not work well. Figure 2.3 show two examples. In the first example, there were 3 major events related to a release of new devices: *new device announcement*, *new device pre-order*, and *new device available*. If we consider all the customer care calls that are related to a new device, we can see clearly there are 3 spikes in the call volume corresponding to the above events. However, the events have little impact on the total call volume, and are difficult to detect using the total call volume. In the second example, 3G network outage occurred in South Florida on the second day of the third week. The anomaly can be detected using the weighted sum of call volume from categories "Technical", "Cannot Make or Receive Calls", "Voice", and "FLP" (Florida/Puerto Rico area), but not from the total call volume. Simply aggregating all calls is insufficient because (i) some events only have impact on a subset of customers and do not lead to significant changes in total volume, and (ii) even for the events that may potentially affect the total volume, the capacity of call centers limits the increase in total volume and makes it difficult to detect. Ideally in this case, we want to detect events before the capacity limit is reached so that we can increase call center capacity temporarily in response to the increased demand.

Another natural approach to detect anomalies is to use PCA. For example, [3] decomposes data into normal and abnormal subspaces using PCA by (i) applying PCA to the testing dataset and examining the projection of testing dataset on each principal axis in order, (ii) assigning the principal axis and all subsequent axes to anomalous subspace as soon as a projection is found to exceed a threshold (e.g., 2x standard deviation from mean), and (iii) detecting spikes in the time series projected onto the anomalous subspace. We observe its precision (i.e., fraction of predicted anomalies that are correct) is not much better than that of a random algorithm, which reports an anomaly by tossing a bias coin of 0.3 (i.e., close to the fraction of time that has anomalies.) PCA performs poorly for several reasons. First, large anomalies can pollute the normal subspace [5] so PCA usually requires anomaly-free data for training [4]. This could be a problem for our dataset because there are average 1.8 anomalies per week in NCCO reports and it is hard to find a clean period for training. Second, determining the threshold for anomalous subspace is an open question [3–5]. Third, PCA is sensitive to noise, which is common in the customer care call dataset.

In general, the appropriate aggregation depends on the types of events. Due to a large number of possible types of events and evolving nature of events, it is infeasible to manually determine the aggregation for each event type in advance.

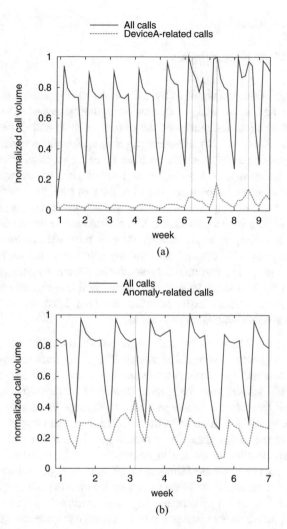

**Fig. 2.3** The total call volume is insufficient to detect events. (**a**) The three vertical green lines indicate the time of 3 major events related to Device-A. (**b**) An anomaly detected using the weighted sum of several categories

The above results call for a new method to automatically learn the mapping from the various input metrics to anomalies. The method should be (i) adaptive to new call labels and events, (ii) highly robust against noise, which is inherent in call data records due to different customers' responses to anomalies and inconsistent call labelings.

## 2.2  Our Approach

### 2.2.1  Problem Formulation

Our main problem can be formulated as follow. We have training and testing traces, where the training traces contain $N$-dimension input timeseries and the ground-truth about when anomalies take place and the testing traces contain new $N$-dimension input timeseries. Our goal is to determine when anomalies occur in the testing traces.

We use regression to approach the problem by casting it as an inference problem of $Ax = b$, where $A(t, i)$ denotes the number of calls of category $i$ at time $t$, $x(i)$ denotes the weight (importance value) of the $i$-th category, $b(t)$ is an indicator whether there is an anomaly. We construct $A$ and $b$ from the training traces to solve for $x$. Then we plug in the estimated $x$ and construct $A$ from the testing traces to predict $b$ for the testing trace. Essentially we view there exists a linear relationship between the categories values and the resulting anomalies, and we try to learn the linear coefficients $x$ that automatically combines different metrics to predict the anomalies. As we will show in Sect. 2.3, a simple linear regression model works well, so we believe the linear assumption is reasonable. There are several significant challenges involved in realizing this scheme.

1. Dynamic $x$: As the categories and events evolve, the relationship between the inputs and the anomalies may also change. Therefore $x$ can change over time.
2. Under-constraints: The number of categories can be much larger than the number of constraints derived from the training traces. So we have an under-constrained problem and there are an infinite number of solutions. Randomly picking one of them gives an equally good fit to the training data, but can give very different accuracy for the testing data. Our ultimate goal is to find $x$ that can accurately predict the anomalies for the testing traces.
3. Over-fitting: Even if we are fortunate to get long enough training traces so that the number of categories is close to or smaller than the number of constraints, the weight estimation is still challenging because the solution that minimizes the fitting error based on the training traces is often not the one that gives the closest fit to the testing data. In other words, there can be over-fitting issues.
4. Scalability: There are thousands of categories or dimensions and thousands of time intervals. The scalability issue further exacerbates when we allow $x$ to change. For example, if there are $K$ different $x$'s, the problem size further grows by a factor of $K$.
5. Varying customer response time: When an anomaly occurs, customers respond to it at different time depending on the impact of anomaly, the customers' own availability, time of day, and day of week. This blurs the relationship between $A$ and $b$.

Below we first develop an approach to address the under-constraints and over-fitting issues while allowing $x$ to change over time (Sect. 2.2.2). Then we reduce the number of unknowns as well as handle the scalability issues by clustering categories and identifying important clustered categories (Sect. 2.2.3). Finally, we use multiple classifiers to enhance the robustness against noise and different customers' response time (Sect. 2.2.4).

### 2.2.2   Our Regression

To address the first challenge, we generalize our formulation to $A_d x_d = b_d$, where $d$ denotes $d$-th day, $A_d(t, i)$ denote the value of $i$-th category from the traces at time $t$ on the $d$-th day, $x_d(i)$ denote the weight of $i$-th variable on the $d$-th day, $b_d(t)$ denote whether there is an anomaly at time $t$ on the $d$-th day, where 1 means anomaly and 0 means no anomaly.

To address the under-constraints and over-fitting issues, we cannot simply minimize the fitting error to the training data. Instead, we also impose additional structures on the solution. First, we expect the weight values $x_d$ to be stable across consecutive days $d$. Second, we expect $x = [x_1 x_2 \ldots x_d]$ exhibits low-rank structure due to the temporal stability in $x$ and the small number of dominant factors that cause the anomalies. Therefore, we try to find $X, U, V$ that minimize the combined objective:

$$o(X, U, V) = \sum_d f(X) + \alpha \cdot g(X) + \beta \cdot h(X, U, V), \tag{2.1}$$

where $f(X)$ is the fitting error, $g(X)$ captures the degree of temporal stability, and $h(X, U, V)$ captures the error in approximating $X$ as a product of two rank $r$ matrices: $U$ and $V$, where $\alpha$ and $\beta$ give the relative weights of temporal stability and low-rank constraints, respectively, and $r$ is the desired low rank. Next we elaborate on each term and how to select weights.

**Fitting Error**   The fitting error is expressed as $f(X) = \sum_d \|A_d x_d - b_d\|_F^2$ where $\| \cdot \|_F$ is the Frobenius norm (with $\|Z\|_F = \sqrt{\sum_{ij} Z(i, j)^2}$ for any matrix $Z$.)

**Incorporating Temporal Stability**   To capture the temporal stability, we introduce a temporal transformation matrix $T$ and define a penalty function as follows:

$$g(X) \stackrel{\triangle}{=} \left\| M * T^T \right\|_F^2, \tag{2.2}$$

where $M = [x_1 x_2 \ldots x_d]$ merges all the column vectors into a matrix and $T^T$ is the transpose of $T$. As in [11], we use a simple temporal transformation matrix to minimize the change in $x$ between two consecutive days: $T = Toeplitz(0, 1, -1)$, which denotes the Toeplitz matrix with central diagonal given by 1, the first upper diagonal given by -1. That is,

$$T = \begin{bmatrix} 1 & -1 & 0 & \dots \\ 0 & 1 & -1 & \ddots \\ 0 & 0 & 1 & -1 & \ddots \\ \vdots & \ddots & \ddots & \ddots \end{bmatrix}. \tag{2.3}$$

**Incorporating Low-Rank Constraints**  Finally, to capture the low-rank nature of weight matrix, we introduce a penalty term function

$$h(X, U, V) \stackrel{\triangle}{=} \left\| M - U * V^T \right\|_F^2 \tag{2.4}$$

where $M = [x_1 x_2 \dots x_d]$, $U$ is a $N \times r$ unknown factor matrix, and $V$ is a $d \times r$ unknown factor matrix, and $r$ is the desired low rank. Minimizing the penalty term ensures $M$ has a good rank-$r$ approximation: $M \approx U * V^T$.

**Selecting Parameters**  To decide the weights $\alpha$ and $\beta$, we use 6-week data as the training set to find the best weights (in terms of evaluation metrics discussed in Sect. 2.3) as follow. First, $\alpha$ and $\beta$ are chosen to make fitting error, temporal stability, and low-rank constraint have similar order of magnitude. Second, we fix $\alpha$, and keep increasing or decreasing $\beta$ by a factor of 10 each time until the performance does not improve. $\beta$ is updated as the one that gives the best performance seen so far. Third, similar to the second step, but this time we fix $\beta$ and alter $\alpha$. The second and third steps are repeated until the performance does not improve for the training traces. It usually takes 5 or fewer iterations. Then we apply the selected parameters to the testing trace for anomaly detection.

Note that in theory, we do not require a separate step of identifying important metrics with the current step of inferring the weights of all metrics as one. However, for efficiency purpose, we separate these two steps in our approach because the step of identifying important metric requires working with all metrics, which are potentially thousands and $L_1$ norm minimization is light-weight enough to work with thousands of metrics.

## 2.2.3  Reducing Categories

The regression problem described in Sect. 2.2.2 has thousands of variables for each day alone and the number further increases with the number of days. This imposes both scalability issue and exacerbates under-constraint issue.

### 2.2.3.1 Clustering Categories

To enhance scalability and minimize the impact of ambiguity and inconsistency in detecting events based on the customer needs categories, we need to cluster relevant categories in advance. It is infeasible to group categories manually because (i) the number of categories changes over time. Categories can be added or depreciated because of the emerging or ending of services/products; (ii) each customer care center can change the usage of categories; and (iii) there are too many categories.

One natural approach is to cluster categories base on the similarity between timeseries. However, the approach does not work. For example, Fig. 2.4 shows timeseries of different categories. We can see that the call volume of category "Plans and Features" increases significantly while that of the other category "Plan" decreases on the same day, because the agents are asked to switch to use "Plans and Features" instead of "Plan". Similarly, we can see the call volume of "Device" decreases while that of "Account" and "Equipment" increases. Although these categories represent the same group of calls, their timeseries are not similar.

We cluster categories based on the similarity of their textual names since agents usually classify calls based on the textual names of categories and different agents may classify similar calls to different categories with similar texts. We treat each category name as a sequence of characters and adopt the Dice's coefficient [9] using bigram model [8], which is widely used in statistical natural language processing to measure string similarity. The Dice's coefficient $s$ for two string $x$ and $y$ using the bigram model is computed as $s = \frac{2 \times n_t}{n_x + n_y}$, where $n_t$ is the number of character the bigrams in both strings, $n_x$ and $n_y$ are the numbers of big-rams in strings $x$ and $y$, respectively. The value of $s$ ranges between 0 and 1. A larger $s$ indicates the two strings are more similar. For example, to calculate the similarity between strings "paid" and "payment", we find the sets of the bigrams as ("pa", "ai", "id") and ("pa", "ay", "ym", "me", "en", "nt"). These sets have 3 and 6 elements, while only 1 element is common. So we have $s = (2 \times 1)/(3 + 6) = 0.22$. By using a threshold of 0.3 for $s$, we cluster the customer categories into 96 at the first level, 354 at the second level, and 1165 at the third level. Our evaluate uses these newly computed categories.

### 2.2.3.2 Identifying Important Categories

Even after clustering, there are still a large number of clustered categories. We use the following three schemes to identify important categories.

**Principal Component Analysis (PCA)** PCA can also be used to identify important categories because it reduces dimensionality of a multivariate dataset by converting possibly correlated variables into linearly uncorrelated variables, which are called principal components. However, using PCA in this context has the same problem as it is used for anomaly detection mentioned in Sect. 2.1. As a result, principal components may not identify the most important categories for anomaly detection.

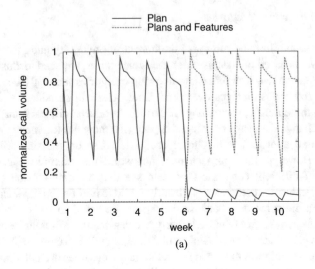

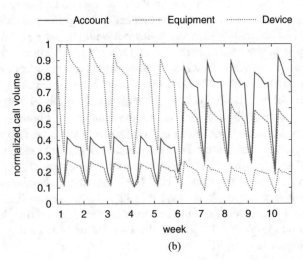

**Fig. 2.4** Dynamic call labels. (**a**) The number of calls labeled as "Plan" and "Plans and Features". (**b**) The number of calls labeled as "Account", "Equipment", and "Device"

For example, the 3G network outage event shown in Fig. 2.3 is dominated by categories "Technical", "Cannot Make or Receive Calls", "Voice", and "FLP". However, the PCA results show that in the top 10% principal components, the coefficients of "Cannot Make or Receive Calls" and "Voice" are small, which indicates they are not considered as important categories in PCA.

$L_2$ **Norm Minimization** Another way of finding important categories is to cast it as an inference problem $Ax = b$, where $A(t, i)$ denotes the value of category $i$ at time $t$, $x(i)$ denotes the weight (importance value) of the $i$-th category, $b(t)$ is an indicator whether there is an anomaly. Different from Sect. 2.2.2, here we just need to filter out unimportant categories instead of determining the precise weights. So here we assume $x$ is constant over time to have fewer unknowns.

We obtain $A$ and $b$ learned from the previous traces. Then we estimate $x$ to best fit the relationship. A common metric for the best fit is $L_2$ *norm minimization*, defined as follows:

$$\min_x \|b - Ax\|_2^2 + \lambda^2 \|x\|_2^2, \qquad (2.5)$$

where $\|x\|_2 = \sqrt{\sum_{k=1..n} \|x_k\|^2}$. It can be efficiently solved using a standard solver for linear least-squares problems. Then we filter out the categories whose weight $x$ is within a threshold, which is set to 0 in our evaluation.

$L_1$ **Norm Minimization**   Another approach is to use $L_1$ *norm minimization*, defined as follow:

$$\min_x \|b - Ax\|_2^2 + \lambda^2 \|x\|_1. \qquad (2.6)$$

where $\|x\|_1 = \sum_{k=1..n} \|x_k\|$. $L_1$ norm minimization is often used in situations where $x$ is *sparse*, i.e., $x$ has only very few large elements and the other elements are all close to 0. This is well suited to our goal of identifying a small number of important factors. As shown in [1], the minimal $L_1$ norm solution often coincides with the sparsest solution for under-determined linear systems. As we will show in Sect. 2.3, $L_1$ norm minimization performs the best since it explicitly maximizes sparsity. As before, we filter out the categories whose weight $x$ is within a threshold, which is set to 0 in our evaluation.

### 2.2.4   Combining Multiple Classifiers

**Need for Multiple Classifiers**   Another important problem is what time scale we should use for anomaly detection. Ideally, we would like to capture all calls triggered by the same anomaly when learning the weight of the metrics. That is, $A$ should include the characteristics of all calls corresponding to that anomaly. However, customers do not respond to an anomaly immediately and sometimes their response time may differ by hours. But simply using a large time window is not a good option since we can no longer detect anomalies in a fine time granularity.

To address both issues, we use a reasonably small bin size: 1 hour, but include calls made in previous $n$ and next $m$ hours as additional features. That is, we use $A_d(t - m, t - (m - 1), \ldots, t - 1, t, t + 1, \ldots, t + n)$, which denotes the values of $N$ categories in the traces from time $t - m$ to time $t + n$. So there are altogether $(m + n + 1) \times N$ features and $x_d$ also now has $(m + n + 1) \times N$ elements, which are the weights of all these features. $b_d(t)$ remains the same as before (i.e., whether there is an anomaly at time $t$).

However it is challenging to select $m$ and $n$ a priori. One set of values may work well on some data but not on others. Therefore we use multiple classifiers, where each classifier uses one set of $m$ and $n$, and then we aggregate the results of all the classifiers. The intuition is that it is more likely to be a real anomaly if lots of classifiers claim so.

**Aggregating Multiple Classifiers** We apply each classifier independently to the testing data and returns a binary timeseries $pb(c, t)$. $pb(c, t) = 1$ denotes that there is an anomaly detected by classifier $c$ at time $t$. $pb(c, t) = 0$ denotes there is no anomaly detected. We aggregate $pb(c, t)$ by assigning a weight $w_c$ to each classifier. We detect an anomaly when $\sum_c w_c pb(c, t) > threshold$. Our evaluation uses a threshold of 0.3.

We calculate $w_c$ for a classifier $c$ by applying 2-fold cross-validation to the training data. The 2-fold cross-validation partitions the training data into two parts. In the first round, it uses the first partition for training and the second partition for testing. Since we know the ground truth in all training data, we can evaluate how the classifier performs in cross-validation by calculating the accuracy (i.e., the fraction of correct prediction) in the second partition. Similarly, in the second round, we use the second partition of the training data for training, use the first partition for testing, and calculate the accuracy in the first partition. Therefore, with the 2-fold cross-validation we can get an average accuracy $a_c$ in training set which gives us an estimate how the classifier may perform. Then the weight of each classifier is assigned as the normalized accuracy: $w_c = a_c / \sum_c a_c$.

## 2.3  Evaluation

We use the following metrics to quantify the accuracy:

$$recall = \frac{tp}{tp + fn} \tag{2.7}$$

$$precision = \frac{tp}{tp + fp} \tag{2.8}$$

where $tp$ is the number of true positives (i.e., correctly detected anomalies), $fp$ is the number of false positives (i.e., incorrectly detected anomalies), and $fn$ is the number of false negatives (i.e., missed anomalies). In addition, we integrate precision and recall into a single metric called *F-score* [10], which is the harmonic mean of precision and recall: $F\text{-}score = \frac{2}{1/precision + 1/recall}$. For all three metrics, larger values indicate higher accuracy. Unless otherwise specified, we use 30 classifiers.

**Identification of Important Features** We first evaluate how different feature selection algorithms impact the performance. We vary the methods of identifying

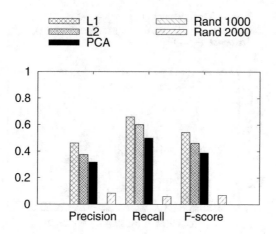

**Fig. 2.5** Varying schemes for features

important categories while using multiple classifiers and temporal/low-rank based regression. We compare PCA, $L_1$ norm minimization, $L_2$ norm minimization, and random selection (e.g., *Rand 1000* and *Rand 2000* randomly select 1000 and 2000 categories). $L_1$ norm selects 612 important categories and $L_2$ norm selects 980 important categories. As shown in Fig. 2.5, $L_1$ norm consistently performs the best. It out-performs $L_2$ norm by 23%, PCA by 45%, Rand 2000 by 454% in terms of precision; out-performs $L_2$ norm by 10%, PCA by 32%, and *Rand 2000* by 1020% in terms of recall. *Rand 1000* selects too few categories and yields close to 0 precision and recall, so its bars are almost invisible from the figure.

**Varying Regression Methods** Next we evaluate the impact of regression methods. We use $L_1$ norm minimization to select important categories and use multiple classifiers in all cases. We compare regression (i) that only uses fitting error as the objective (*Fit*), (ii) that uses fitting error and low rank (*Fit+LR*), (iii) that uses fitting error and temporal stability (*Fit+Temp*), (iv) that uses fitting error, temporal stability, and low rank (*Fit+Temp+LR*). In addition, as a baseline, we use random selection (Rand 0.3) that randomly determines if a given interval has an anomaly with a probability of 0.3 since around 30% of time intervals have anomalies. As shown in Fig. 2.6, Fit+Temp+LR yields the highest accuracy: it out-performs Random, Fit, Fit+LR, Fit+Temp by 823%, 64%, 32%, 6%, respectively, in terms of F-score.

**Using Multiple Classifiers** We evaluate how the number of classifiers affects the performance. As shown in Fig. 2.7, leveraging more classifiers can generally improves the accuracy, as we would expect. The improvement increases significantly initially and then tapers off. Since the computation cost increases with the number of classifiers, we use 30 classifiers as the default to trade off the benefit and cost.

**Varying the Sizes of Training Sets** We use the previous $n$ days as the training set and detect anomalies on the next day. As shown in Fig. 2.8, when $n$ increases from

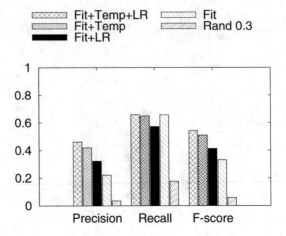

**Fig. 2.6** Varying regression methods

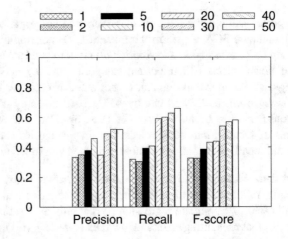

**Fig. 2.7** Varying the number of classifiers

7 days to 42 days, the recall increases from 46% to 79%, but when $n$ increases from 42 days to 63 days, the recall drops to 51%. Similarly, when $n$ increases from 7 days to 21 days, the precision increases from 44% to 61%, but a further increase in $n$ to 63-day reduces the precision to 32%. It is because when training set is larger, we have more constraints to find a better solution for Eq. (2.1) and therefore a higher accuracy. However, as the training set further increases and includes older dataset, the performance may degrade due to the evolving nature of the dataset. We plan to place higher weights to the constraints learned from more recent dataset to further improve the accuracy and robustness in the future.

**Varying Ratios of Weight**  Let $x(i, t)$ denote the weight (importance value) of the $i$-th category at time $t$. The change ratio of weight is $\sum_t \sum_i (x(i, t) - x(i, t -$

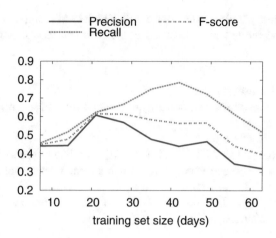

**Fig. 2.8** Varying the size of training set

$1))/x(i, t - 1)$ when $x(i, t - 1)$ is not 0. We see 84% of the changes is within 10%, and 2.5% of the changes is larger than 100%. This suggests $x$ has significant temporal stability, but it can also adapt to different values in order to detect different types of anomalies across two consecutive days.

## 2.4 External Data: Social Media

So far, we focus on event detection using the call records, which are the direct feedbacks from customers. However, it is hard to understand the nature of detected events from customer calls due to the following reasons: (i) Categories from call records only have limited text information to describe the issues behind the calls. (ii) Locations of the called customers can be inferred by the area codes of their phone numbers. However, the coverage of each area code is not uniform. Moreover, the customers may use the area codes from their previous locations.

**Benefits of Using Twitter** To better understand the detected events, we leverage Twitter social media as an indirect channel to understand customers' experience of the service. There are mainly three reasons that make Twitter social media an attractive data source. First, Twitter data is massive; as of March 2012, Twitter has 500 million registered users [7]. Many people share their experiences of the services and products they are using. Second, user feedbacks are coming in near real-time. Compared with the efforts to report issues through customer calls, it is much easier to express their experiences through *tweets*. Third, tweets may have richer context information than customer calls. They have many features which help to understand the various events.

**Data Description**   Twitter data comes with a variety of features, which will help us to understand the nature of detected events. We use the following features of tweets:

- Timestamp;
- Text of original tweets
- Text of normalized tweets, which are the tweets converted into a standard format to ease processing;
- Username: a tweet's author;
- Twitter specific features: URLs, retweets (RT), #hashtags used to mark keywords or topics in a tweet, and @mentions, which are the tweets containing "@user-name" anywhere in the text;
- Location (city, state, latitude, longitude): Locations can come from the tweets themselves or Foursquare [2].

**Understanding Detected Events**   Once the anomalies are detected, we can leverage various features of tweets, such as keywords and locations, to summarize events as follow. We first find tweets that are related to customers experience by selecting tweets with hashtag #XXXFAIL or #XXXSUCK during the entire period, where XXX denotes the name of the provider. Then we try to identify important keywords that appear in the anomaly period. We use the metric, called Term Frequency - Inverse Document Frequency (TF-IDF) [6], to quantify the importance of a keyword. TF-IDF is defined as the number of occurrences of word-level 1-grams and 2-grams during the period of the anomaly (in the selected tweets) divided by the number of occurrences in the entire period (in the selected tweets). The intuition is that a keyword that appears frequently only in the anomaly period but not universally frequently is important. Table 2.2 shows some examples of events summarized using this approach.

To locate the impacted regions of the given anomaly, we gather the authors of the collected tweets, which contains n-grams with high TF-IDF scores during the

**Table 2.2** Examples of detected anomalies with the summary

| Event | Location | Event summary by TF-IDF |
|---|---|---|
| 3G network outage | New York, NY | Service, outage, nyc, calls, ny, morning, service |
| Outage due to an earthquake | East Coast | #earthquake, working, wireless, service, nyc, apparently, new, york |
| 3G network outage | Miami, FL | Outage, south, service, issue, broward (a county in FL), key, west, equipment, Florida |
| Internet service outage | Bay Area | serviceU, bay, outage, service, Internet, area, support, #fail |
| New device release | Nationwide | iphone, sprint, verizon, apple, 4s, android, accessibility |
| New device release | Nationwide | 4s, apple, iphone, #iphone4s, pre-order, order, site, #apple, store |

**Fig. 2.9** Tweet locations of an detected event

time frame. Then we check the locations of the authors to get the impacted regions. Figure 2.9 shows an example of the tweet locations for the anomaly occurred at Miami, FL. Although some users may tweet about network incidents from other locations, as long as we see multiple tweets in a region, we can correctly locate the anomaly.

## 2.5 Conclusion

We develop a systematic method to automatically detect anomalies in a cellular network using the customer care call data. Our approach scales to a large number of features in the data and is robust to noise. Using evaluation based on the call records collected from a large cellular provider in US, we show that our method can achieve 68% recall and 86% accuracy, much better than the existing schemes. Moreover, we show that social media can be used as a complementary source to get higher confidence on the detected anomalies and to summarize the user feedbacks to anomalies with text and location information.

## References

1. D. Donoho, For most large underdetermined systems of linear equations, the minimal L1-norm near-solution approximates the sparsest near-solution (2004). http://www-stat.stanford.edu/~donoho/Reports/2004/l1l0approx.pdf
2. Foursquare API (2021). http://developer.foursquare.com/docs/
3. A. Lakhina, M. Crovella, C. Diot, Diagnosing network-wide traffic anomalies, in *Proc. of ACM SIGCOMM*, 2004
4. A. Lakhina, K. Papagiannaki, M. Crovella, C. Diot, E.D. Kolaczyk, N. Taft, Structural analysis of network traffic flows, in *Proc. of ACM SIGMETRICS*, 2004

5. H. Ringberg, A. Soule, J. Rexford, C. Diot,  Sensitivity of PCA for traffic anomaly detection, in *Proc. of ACM SIGMETRICS*, 2007
6. TF-IDF (2022). http://en.wikipedia.org/wiki/Tf*idf
7. Twitter to surpass 500 million registered users on Wednesday (2022). http://www.mediabistro.com/alltwitter/500-million-registered-users_b18842
8. Wikipedia, Bigram (2022). http://en.wikipedia.org/wiki/Bigram
9. Wikipedia, Dice's coefficient (2022). http://en.wikipedia.org/wiki/Dice's_coefficient
10. Wikipedia,. F1 score (2022). http://en.wikipedia.org/wiki/F1_score
11. Y. Zhang, M. Roughan, W. Willinger, L. Qiu, Spatio-temporal compressive sensing and internet traffic matrices. ACM SIGCOMM Comput. Commun. Rev. **39**(4), 267 (2009)

# Chapter 3
# Limitation of Compressive Sensing

Real-world data are complicated and heterogeneous, and often violate the low-rank assumption required by existing compressive sensing techniques. Such violation significantly reduces the applicability and effectiveness of existing compressive sensing approaches. It is important to understand reasons behind the violation to design methods and mitigate the impact. In this chapter, we analyze a wide range of real-world traces in order to reveal the factors that contribute to variability of the real-world data.

## 3.1 Datasets

Table 3.1 lists the network matrices used for our study. We obtained 3G traces from 472 base stations in a downtown area of a major city with a population of 8.7 million people in China during Oct. 2010. We generate traffic matrices by computing the aggregated traffic every 10 minutes. $M(i, t)$ represents the total traffic volume to and from base station $i$ during time interval $t$, where $t$ is a 10-minute interval.

We also got WiFi traffic from a large university in China, and generated traffic matrices based on the traffic collected at the 50 most loaded access points (APs) on Jan. 4, 2013. $M(i, t)$ denotes the total traffic to/from AP $i$ during the $t$-th 10-minute time interval.

In addition to wireless traffic matrices, we also use traffic matrices from Abilene (Internet2) [19] and GÉANT [21], which are standard traffic matrices used in several previous studies (e.g., [13, 14, 23, 24]) and useful for comparison. Abilene traces report total end-to-end traffic between all pairs in a 11-node network every 10 minutes. GÉANT traces report total traffic between all pairs in a 23-node network every 15 minutes.

For diversity, in addition to traffic matrices, we also obtain signal strength and expected transmission time (ETT). We use the following SNR and RSS matrices: (a)

© The Author(s), under exclusive license to Springer Nature Switzerland AG 2022
G. Xue et al., *Robust Network Compressive Sensing*, SpringerBriefs in Computer
Science, https://doi.org/10.1007/978-3-031-16829-1_3

**Table 3.1** Network datasets

| Network | Date | Duration | Resolution | Size |
|---|---|---|---|---|
| 3G traffic | Nov. 2010 | 1 day | 10 min. | $472 \times 144$ |
| WiFi traffic | Jan. 2013 | 1 day | 10 min. | $50 \times 118$ |
| Abilene traffic [19] | Apr. 2003 | 1 week | 10 min. | $121 \times 1008$ |
| GÉANT traffic [21] | Apr. 2005 | 1 week | 15 min. | $529 \times 672$ |
| 1 channel CSI | Feb. 2009 | 15 min. | 1 frame | $90 \times 9000$ |
| multi. channels CSI | Feb. 2014 | 15 min. | 1 frame | $270 \times 5000$ |
| Cister RSSI [17] | Nov. 2010 | 4 hours | 1 frame | $16 \times 10,000$ |
| CU RSSI [3] | Aug. 2007 | 500 frames | 1 frame | $895 \times 500$ |
| UMich RSS [22] | April 2006 | 30 min. | 1 frame | $182 \times 3127$ |
| UCSB Meshnet [20] | April. 2006 | 3 days | 1 min. | $425 \times 1527$ |

our 1 channel CSI traces, (b) our multi-channel CSI traces, (c) Cister RSSI traces, (d) CU RSSI traces, and (e) UMich RSS. We collected (a) by having a moving desktop transmit back-to-back to another desktop and letting the receiving desktop record the SNR across all OFDM subcarriers over 15 minutes using the Intel Wi-Fi Link 5300 (iwl5300) adapter. The modified driver [1] reports the channel matrices for 30 subcarrier groups in a 20MHz channel, which is about one group for every two subcarriers according to the standard [15]. The sender sends 1000-byte frames using MCS 0 at a transmission power of 15 dBm. Since MCS0 has 1 stream and the receiver has 3 antennas, the NIC reports CSI as a $90 \times 1$ matrix for each frame. We collect (b) on channels 36, 40, 44, 48, 149, 153, 157, 161, and 165 at 5GHz. The transmitter starts from channel 36, and sends 10 packets before switching to the next channel. The receiver synchronizes with the transmitter and also cycle through the 9 channels, which yields $270 \times 1$ matrix for each frame. In addition, we use Cister RSSI traces [17], CU RSSI traces [3], and UMich RSS traces [22], all of which are publicly available at CRAWDAD [5]. $M(f, i)$ in Cister traces denotes RSSI on IEEE 802.15.4 channel $f$ in the $i$-th frame, $M(l, i)$ in CU traces denotes RSSI of the $i$-th frame at location $l$, and $M(s, i)$ in UMich-RSS trace denotes the RSS measurement received by the $s$-th sensor pair in the $i$-th packet. We also use ETT traces from UCSB Meshnet [20], which contains the ETT of every links in a 20-node mesh network. We generate the ETT matrix $M(l, t)$, where $M(l, t)$ denotes the ETT of the link $l$ during the $t$-th 10-second window.

The last trace in Table 3.1 is Foursquare check-ins dataset obtained from [4], which was collected from Foursquare location service across 3927 locations. They used the location crawler to collect information from 225,098 users, altogether 22,506,721 unique check-ins. More than 53% of the check-ins are from Foursquare, and most of the other check-ins are from Twitter's mobile applications and other location sharing services, such as Gowalla, Echofon, and Gravity. As most of the places have a very few check-ins, we have aggregated check-ins on a 2-hourly basis in 6 months over 3927 unique places across the globe. $M(l, t)$ in Foursquare traces denotes the number of check-ins at location $l$ at time interval $t$.

**Table 3.2** Activity datasets. The listed matrix sizes include all users. For some analyses, we use per-user matrices, which is roughly $1/NumUsers$ of the above matrix sizes

| Dataset | Date | #Users | #Activities | Size |
|---|---|---|---|---|
| Waist accelerometer [11] | Oct. 2010 | 30 | 6 | $10,299 \times 561$ |
| Wrist accelerometer [8] | Feb. 2014 | 16 | 14 | $429,177 \times 3$ |
| Chest accelerometer [2] | Feb. 2014 | 15 | 7 | $1,926,881 \times 5$ |
| PAMAP2 [18] | June 2012 | 9 | 18 | $3,850,505 \times 52$ |
| EMG [6] | March 2013 | 8 | 14 | $17,920,000 \times 8$ |
| P300 [7] | March 2007 | 9 | 6 images | $2,949,120 \times 34$ |

We also use several activity and brain wave traces, as shown in Table 3.2. The first trace [11] was collected from 30 volunteers performing six activities: walking, walking upstairs, walking downstairs, sitting, standing, and lying down while wearing smartphones on their waists and recording the accelerometer and gyroscope data at 50 Hz sampling rate. 561 features were extracted in time and frequency domain for each window. Example features include mean, variance, maximum, minimum, energy, entropy, frequency components of the accelerometer and gyroscope readings.

The second trace is labeled accelerometer data collected for Activities for Daily Living (ADL) recognition using a wrist-worn tri-axial accelerometer [8]. The sampling rate is 32 Hz. There are 14 activities performed by 16 volunteers: brushing teeth, climbing stairs, combing hair, descending stairs, drinking glass, eating meat, eating soup, getup from bed, lie down on bed, pouring water, sitting down on chair, standing up on chair, using telephone, walking.

The chest-mounted accelerometer dataset [2] contains uncalibrated accelerometer data from 15 participants performing 7 activities: working at Computer, standing up, standing, walking, going up and down Stairs, walking and talking with someone, talking while standing at 52 Hz sampling rate.

The PAMAP2 Physical Activity Monitoring dataset [18] contains data from 3 inertial measurement units and a heart rate monitor worn by 9 users performing 18 activities (e.g., walking, cycling, playing soccer).

The EMG dataset [6] contains 8 participants (20–35 year old) carrying out 14 finger movements (i.e., 12 finger pressures and 2 finger points). The participants were asked to perform each movement six times, and hold them for 5 seconds in each trial [12]. Four of six repeated movements were labeled. We use these labeled traces.

The P300 EEG dataset [7] was recorded from 9 users watching a laptop screen, which displayed six images including a television, a telephone, a lamp, a door, a window, or a radio. The images were flashed in random sequences, one image at a time. One of the six images was selected as the target image, and users were asked to count how often the target image was flashed. Different users are assigned different target images. The goal is to analyze brain waves collected when a user is watching an image to automatically identify if a user is looking at a target image. The EEG samples data at 2048 Hz from 32 electrodes placed at the standard positions in the 10–20 international system [9].

## 3.2  Analysis

In this section, we use the datasets listed in Tables 3.1 and 3.2 to reveal the factors that contribute to variability of the real-world data.

### 3.2.1  Anomalies Increase Ranks

For each network matrix, we mean center each row (i.e., subtract from each row its mean value). We then apply singular value decomposition (SVD) to examine if the mean-centered matrix has a good low-rank approximation. The metric we use is the fraction of total variance captured by the top $K$ singular values, i.e., $\left(\sum_{i=1}^{K} s_i^2\right) / \left(\sum_i s_i^2\right)$, where $s_i$ is the $i$-th largest singular value and $\left(\sum_i s_i^2\right)$ gives the total variance of the mean-centered coordinate matrix. Note that $1 - \left(\sum_{i=1}^{K} s_i^2\right) / \left(\sum_i s_i^2\right)$ is the relative approximation error of the best rank-$K$ approximation with respect to the squared Frobenius norm.

Figure 3.1 plots the fraction of total variance captured by the top $K$ singular values for different traces. As it shows, from low to high, UCSB Meshnet, GÉANT, multi-channel CSI, UMich RSS, RON, 1-channel CSI, CU RSSI, Abilene, WiFi, 3G, and Cister RSSI matrices take 7.5%, 20.8%, 22.0%, 23.9%, 47.2%, 48.9%, 55.8%, 57.0%, 58.0%, 68.1%, and 81.0% singular values to capture 90% variance, respectively. Therefore, only UCSB Meshnet, GÉANT, multi-channel CSI, and UMich RSS are close to low rank.

Next we inject anomalies to see how it affects the results. We inject anomalies to a portion of the entries in the original matrices. Following the standard anomaly injection method used in existing work [10, 13, 16], we first use exponential weighted moving average (EWMA) to predict the future entries based on their past

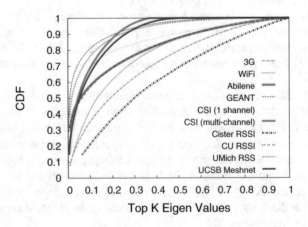

**Fig. 3.1**  CDF of ranks

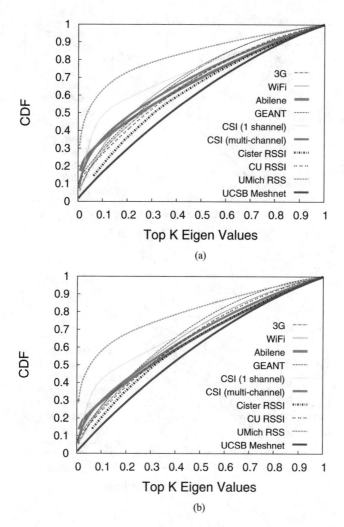

**Fig. 3.2** Ranks under anomalies in traffic matrices with $s = 0.5$. (**a**) 5% anomalies. (**b**) 10% anomalies

values (i.e., $y = \alpha x + (1 - \alpha)y$, where $\alpha = 0.8$ in our evaluation) and use the maximum difference between the actual and predicted value as the anomaly size to be injected. We vary the fraction of entries to inject anomalies from 5% to 10%, and also scale the anomaly size by $s$, which is 0.5 or 1.

As shown in Figs. 3.2 and 3.3, when we inject more anomalies or larger anomalies, more singular values are required in order to capture the variance of the matrices. This trend is consistent across all traces. For example, as shown in Fig. 3.2, when we inject 5% and 10% anomalies with s=0.5, it takes 60.9% and 67.6% singular values to capture 90% variance in UMich Meshnet, 71.0% and

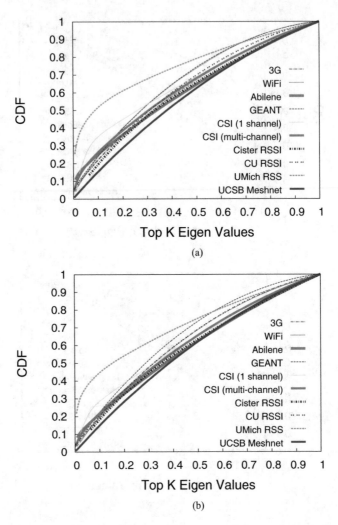

**Fig. 3.3** Ranks under anomalies in traffic matrices with $s = 1$. (**a**) 5% anomalies. (**b**) 10% anomalies

75.8% in WiFi trace, and 75.7% and 80.0% in 1-channel CSI trace. As shown in Fig. 3.3, when we inject 5% and 10% anomalies with s=1, the corresponding numbers are 72.5% and 76.9% in UMich Meshnet, 72.0% and 78.3% in WiFi trace, and 81.7% and 84.1% in 1-channel CSI trace. Similarly, the other matrices exhibit the same trend: it takes 79.3% and 81.8% singular values for Abilene to capture 90% variance, when we inject 5% and 10% anomalies with an average size of 0.1, respectively. The corresponding numbers rose to 85.1% and 86.0% when we inject 5% and 10% anomalies with an average size of 0.5, respectively.

### 3.2.2 Lack of Synchronization Increases Ranks

We compute the rank of the matrices formed by the traces in Table 3.1, where the rank is defined as the minimum number of eigenvalues required to capture at least 80% energy. We compare the ranks of the original traces (*original*), the ranks of the traces after shifting each row in the matrix by a random offset (*random*), which serves as the worst-case performance, and the ranks of the traces after going through our data-driven synchronization (*sync*). Note that a matrix after applying synchronization or random shift is no longer a matrix as shown in Fig. 5.3, so we take overlapping columns across all the rows to compute the rank. For fair comparison, we ensure the same matrices size across all three schemes (e.g., extracting the same number of columns from the original matrix to compute its rank).

Figure 3.4 summarizes the results. We make several observations. First, the ranks after synchronization are lowest. They are 0–36% lower than the original ranks, and 6–65% lower than the ranks under random shifts. It is interesting that the rank can often be reduced over the original traces. This is because of time zone difference (e.g., nodes in Four Square span different time zones), clock synchronization error (e.g., nodes in WiFi and UMichigan traces are on the same campuses, but have synchronization error), and propagation time of network events (e.g., likely in 3G due to large geographic area the nodes span even though they are in the same time zone). Second, the reduction varies significantly across traces. For example, the rank reduces by 65% over random offset in multi-channel CSI, and reduces by 36-38% over both random offset and original traces in CU RSSI, whereas the rank changes little in Cister. Third, the ranks after random shifts are highest due to lack of synchronization (e.g., the rank increases by 65% in multi-channel CSI with a random shift).

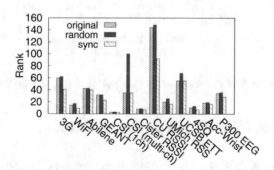

**Fig. 3.4**  # top singular values that capture 80% energy

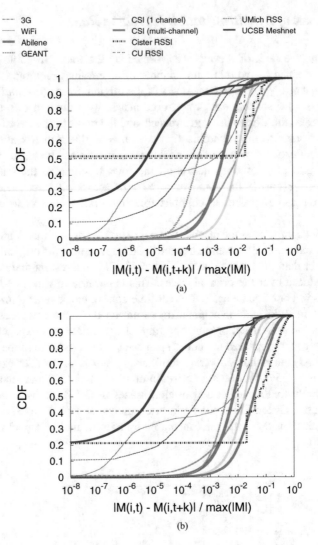

**Fig. 3.5** CDF of normalized difference between i-th and i+k-th time slot. (**a**) $k = 1$. (**b**) $k = 10$

### 3.2.3  Temporal Stability Varies Across Traces

Figure 3.5 plots the CDF of normalized temporal variation (i.e., $\frac{x(i)-x(i-t)}{max(x(i))}$) across different traces. As it shows, different traces exhibit varying degrees of temporal stability. For example, 3G and Cister RSSI have high variation: the two adjacent entries differ by 6.1%–9.8% in 90% cases, and 10 time-slot apart entries differ by 16.7%–36.2% in 90%. In comparison, UMich Meshnet and GÉANT have low variation, where the adjacent entries differ by 0.3%–0.5% in 90% cases and 10 time-slot apart entries differ by 1.0%–2.4% in 90%. The other traces are in between.

## 3.3 Conclusion

The major findings from the analysis include: (i) Not all real network matrices are low rank. (ii) Adding anomalies further increases the rank. (iii) The lack of synchronization in time domain and frequency domain increases the rank. (iv) Temporal stability varies substantially across different traces. These findings motivate us to develop a general compressive sensing framework to support diverse matrices that may not be low rank, exhibit different degrees of temporal stability, and may even contain anomalies. These results also highlight the importance of synchronization. Synchronizing the traces reduces the ranks and increases effectiveness of compressive sensing, which assumes low-rank matrices. Moreover, it increases the spatial locality in the data, which helps segmentation and activity recognition.

## References

1. 802.11n channel measurement tool (2022). http://ils.intel-research.net/projects/80211n-channel-measurement-tool
2. Activity recognition from single chest-mounted accelerometer data set (2022). https://archive.ics.uci.edu/ml/machine-learning-databases/00287/
3. K. Bauer, D. McCoy, D. Grunwald, D.C. Sicker, CRAWDAD data set CU/RSSI (v. 2009-06-15). http://crawdad.cs.dartmouth.edu/cu/rssi/
4. Z. Cheng, J. Caverlee, K. Lee, D.Z. Sui, Exploring millions of footprints in location sharing services, in *Proc. of ICWSM*, 2011
5. Crawdad (2022). http://crawdad.cs.dartmouth.edu
6. Dataset for identifying individual and combined fingers' pressure on a steering wheel (2022). http://www.rami-khushaba.com/electromyogram-emg-repository.html
7. Dataset of an efficient p300-based brain-computer interface for disabled subjects (2022). http://mmspg.epfl.ch/cms/page-58322.html
8. Dataset for adl recognition with wrist-worn accelerometer data set (2022). https://archive.ics.uci.edu/ml/datasets/Dataset+for+ADL+Recognition+with+Wrist-worn+Accelerometer
9. U. Hoffmann, J.M. Vesin, T. Ebrahimi, K. Diserens, An efficient p300-based brain-computer interface for disabled subjects. J. Neurosci. Methods **167**(1), 115–125 (2008)
10. L. Huang, X. Nguyen, M. Garofalakis, J.M. Hellerstein, Communication-efficient online detection of network-wide anomalies, in *Proc. of IEEE INFOCOM*, 2007
11. Human activity recognition using smartphones data set (2022). https://archive.ics.uci.edu/ml/datasets/Human+Activity+Recognition+Using+Smartphones
12. R.N. Khushaba, S. Kodagoda, D. Liu, G. Dissanayake, Muscle computer interfaces for driver distraction reduction. Comput. Methods Programs Biomed. **110**(2), 137–149 (2013)
13. A. Lakhina, M. Crovella, C. Diot, Diagnosing network-wide traffic anomalies, in *Proc. of ACM SIGCOMM*, 2004
14. A. Lakhina, K. Papagiannaki, M. Crovella, C. Diot, E.D. Kolaczyk, N. Taft, Structural analysis of network traffic flows, in *Proc. of ACM SIGMETRICS*, 2004
15. LAN/MAN Standards Commmittee of the IEEE Computer Society. Part 11: Wireless LAN Medium Access Control and Physical Layer (PHY) Specifications . *IEEE Standard 802.11*, 2009. http://standards.ieee.org/getieee802/download/802.11n-2009.pdf
16. S. Nagaraja, V. Jalaparti, M. Caesar, N. Borisov, P3ca: private anomaly detection across ISP networks, in *Proc. of PETS*, vol. 6794 (Springer, New York, 2011)

17. C. Noda, S. Prabh, M. Alves, T. Voigt, C.A. Boano, CRAWDAD data set Cister/RSSI (v. 2012-05-17). http://crawdad.cs.dartmouth.edu/cister/rssi/

18. Pamap2 physical activity monitoring data set (2022). https://archive.ics.uci.edu/ml/datasets/PAMAP2+Physical+Activity+Monitoring

19. The Abilene Observatory Data Collections (2022). http://abilene.internet2.edu/observatory/data-collections.html

20. UCSB meshnet (2022). http://www.crawdad.org/ucsb/meshnet/

21. S. Uhlig, B. Quoitin, S. Balon, J. Lepropre, Providing public intradomain traffic matrices to the research community. ACM SIGCOMM Comput. Commun. Rev. **36**(1), 83–86 (2006)

22. UMich RSS (2022). http://crawdad.cs.dartmouth.edu/umich/rss/

23. Y. Zhang, Z. Ge, A. Greenberg, M. Roughan, Network anomography, in *Proc. of ACM IMC*, 2005

24. Y. Zhang, M. Roughan, W. Willinger, L. Qiu, Spatio-temporal compressive sensing and internet traffic matrices. ACM SIGCOMM Comput. Commun. Rev. **39**(4), 267 (2009)

# Chapter 4
# Robust Compressive Sensing

In this chapter, we develop LENS decomposition, a novel technique to accurately decompose network data represented in the form of a matrix into a low-rank matrix, a sparse anomaly matrix, an error term, and a small noise matrix. This decomposition naturally reflects the inherent structures of real-world data and is more general than existing compressive sensing techniques by removing the low-rank assumption and explicitly supporting anomalies.

## 4.1 LENS Decomposition

In this section, we first present LENS decomposition framework. Next we develop an alternating direction method for solving the decomposition problem. Then we describe how to set the parameters.

### 4.1.1 LENS Decomposition Framework

**Overview** There are many factors that contribute to real network matrices, including measurement errors, anomalies and inherent noise. To capture this insight, we decompose the original matrix into a Low-rank component, an Error component, a Noise component, and a Sparse anomaly component (hence the acronym LENS decomposition). This is motivated by the following observations:

- Low-rank component: Network matrices often exhibit significant redundancy. A concrete form of redundancy is that the network matrix of interest can be well approximated by low-rank matrices. For example, TM estimation makes use of the gravity model [16], which is essentially a rank-1 approximation to matrices. Reference [11] uses low-rank matrices for localization.

G. Xue et al., *Robust Network Compressive Sensing*, SpringerBriefs in Computer Science, https://doi.org/10.1007/978-3-031-16829-1_4

- Sparse component: anomalies are common in large network dataset. Anomalies may arise from a number of factors. For example, traffic anomalies may be caused by problems ranging from security threats (e.g., Distributed Denial of Service (DDoS) attacks and network worms) to unusual traffic events (e.g., flash crowds), to vendor implementation bugs, and to network misconfigurations. Such anomalies are typically not known a priori and are sparse [4, 9].
  Note that there can be systematic effects that are only sparse after some transformation (e.g., wavelet transform). For example, a major level shift may result in persistent changes in the original data (and is thus not sparse). But after wavelet transform (or simple temporal differencing), it becomes sparse.
- Error and artifacts: The measurement and data collection procedure may also introduce artifacts. For example, a SNMP traffic counter may wrap around, resulting in negative measurements. One can always try his best to apply domain knowledge to filter out obvious errors and artifacts (e.g., missing data or negative traffic measurements). However, in general it is difficult to filter out all such artifacts. The advantage of considering both anomalies and errors jointly is that the parts that cannot be filtered can get absorbed by the sparse components.
- Noise. Noise is universal, making clean mathematical models approximate in practice. For example, real-world network matrices are typically only approximately low-rank as opposed to exactly low-rank.

Therefore a natural approach is to consider the original dataset as a mixture of all these effects. It is useful if one can decompose the original matrix into individual components, each component capturing one major effect.

**Basic Formulation** The basic LENS decomposition decomposes an original $m \times n$ data matrix $D$ into a low-rank matrix $X$, a sparse anomaly matrix $Y$, a noise matrix $Z$, and an error matrix $W$. This is achieved by solving the following *convex* optimization problem:

$$\text{minimize:}\quad \alpha\|X\|_* + \beta\|Y\|_1 + \frac{1}{2\sigma}\|Z\|_F^2,$$
$$\text{subject to:}\quad X + Y + Z + W = D,$$
$$E.*W = W. \tag{4.1}$$

where $X$ is a low-rank component, $Y$ is a sparse anomaly component, $Z$ is a dense noise term, and $E$ is a binary error indicator matrix such that $E[i, j] = 1$ if and only if entry $D[i, j]$ is erroneous or missing, and $W$ is an arbitrary error component with $W[i, j] \neq 0$ only when $E[i, j] = 1$ (thus $E.*W = W$, where $.*$ is an element-wise multiplication operator). Since $W$ fully captures the erroneous or missing values, we can set $D[i, j] = 0$ whenever $E[i, j] = 1$ without loss of generality. The constraint enforces $D$ to be the sum of $X$, $Y$, and $Z$ when $D$ is neither missing nor has errors (since $E[i, j].*W[i, j] = 0$ in this case), while imposing no constraint when $D$ is missing or has error (since $E[i, j].*W[i, j] = W[i, j]$ allows $W[i, j]$ to take an arbitrary value to satisfy $X + Y + Z + W = D$).

The optimization objective has the following three components:

- $\|X\|_*$ is the nuclear norm [12, 13] of matrix $X$, which penalizes against high rank of $X$ and can be computed as the total sum of $X$'s singular values.
- $\|Y\|_1$ is the $\ell_1$-norm of $Y$, which penalizes against lack of sparsity in $Y$ and can be computed as $\|Y\|_1 = \sum_{i,j} |Y[i,j]|$.
- $\|Z\|_F^2$ is the squared Frobenius norm of matrix $Z$, which penalizes against large entries in the noise matrix $Z$ and can be computed as $\|Z\|_F^2 = \sum_{i,j} Z[i,j]^2$.

The weights $\alpha$, $\beta$ and $\frac{1}{2\sigma}$ balance the conflicting goals to simultaneously minimize $\|X\|_*$, $\|Y\|_1$ and $\|Z\|_F^2$. We describe how to choose these weights in Sect. 4.1.3.

**Generalized Formulation**     Below we generalize both the constraints and the optimization objective of the basic formulation in Eq. (4.1) to accommodate rich requirements in the analysis of real-world network matrices.

First, the matrix of interest may not always be directly observable, but its linear transform can be observed though subject to missing data, measurement errors, and anomalies. For example, end-to-end traffic matrices $X$ are often not directly observed, and what can be observed are link load $D$. $X$ and $D$ follow $AX = D$, where $A$ is a binary routing matrix: $A(i, j) = 1$ if link $i$ is used to route traffic for the $j$-th end-to-end flow, and $A(i, j) = 0$ otherwise. We generalize the constraints in Eq. (4.1) to cope with such measurement requirements:

$$AX + BY + CZ + W = D \qquad (4.2)$$

Here $A$ may capture tomographic constraints that linearly relate direct and indirect measurements (e.g., $A$ is a routing matrix in the traffic matrices). $B$ may represent an over-complete anomaly profile matrix. If we do not know which matrix entries may have anomalies, we can simply set $B$ to the identity matrix $I$. It is also possible to set $B = A$ if we are interested in capturing anomalies in $X$. Without prior knowledge, we set $C$ to be the identity matrix.

Prior research on network inference and compressive sensing highlights the importance of incorporating domain knowledge about the structure of the underlying data. To capture domain knowledge, we introduce one or more penalty terms into the optimization objective: $\sum_{k=1}^{K} \|P_k X Q_k^T - R_k\|_F^2$, where $K$ is the number of penalty terms. We also introduce a weight $\gamma$ to capture our confidence in such knowledge.

Examples of domain knowledge include temporal stability constraints, spatial locality constraints, and initial estimation of $X$ (e.g., [16] derives initial traffic matrices using the gravity model [7]). Temporal and spatial locality are common in network data [4, 6, 14]. Such domain knowledge is especially helpful when there are many missing entries, making the problem severely under-constrained.

Consider a few simple cases. First, when $k = 1$, $P_1$ is an identity matrix $I$, $R_1$ is a zero vector, we can set $Q_1 = Toeplitz(0, 1, -1)$, which denotes the Toeplitz matrix with central diagonal given by ones, the first upper diagonal given by negative one, i.e.,

$$Q = \begin{bmatrix} 1 & -1 & 0 & 0 & \dots \\ 0 & 1 & -1 & 0 & \ddots \\ 0 & 0 & 1 & -1 & \ddots \\ \vdots & \ddots & \ddots & \ddots & \ddots \end{bmatrix}. \tag{4.3}$$

$Q^T$ denotes the transpose of matrix $Q$. $P_1 X Q_1^T$ captures the differences between two temporally adjacent elements in $X$. Minimizing $\|P_1 X Q_1^T - R_1\|_F^2 = \|P_1 X Q_1^T\|_F^2$ reflects the goal of making $X$ temporally stable. For simplicity, this is what we use for our evaluation. In general, one can use similar constraints to capture other temporal locality patterns during different periods (e.g., seasonal patterns or diurnal patterns).

Next we consider the spatial locality, which is represented by $P$. If $k = 1$, $R_1 = 0$, $Q_1$ is an identity matrix $I$, we can set $P_1$ to reflect the spatial locality. For example, if two adjacent elements in the matrix have similar values, we can set $P = Toeplitz(0, 1, -1)$. Similarly, if different parts of the matrix have different spatial locality patterns, we can use different $P$'s to capture these spatial locality patterns. For simplicity, our evaluation considers only temporal stability, which is well known to exist in different networks. We plan to incorporate spatial locality in the future.

Finally, if we have good initial estimate of $X_{init}$ (e.g., [16] uses the gravity model to derive the initial TM), we can leverage this domain knowledge by minimizing $\|X - X_{init}\|$ (i.e., $R_1 = X_{init}$). This term can be further combined with spatial and/or temporal locality to produce richer constraints.

Putting everything together, the general formulation is:

$$\text{minimize: } \alpha\|X\|_* + \beta\|Y\|_1 + \frac{1}{2\sigma}\|Z\|_F^2 + \frac{\gamma}{2\sigma}\sum_{k=1}^{K}\|P_k X Q_k^T - R_k\|_F^2,$$

$$\text{subject to: } AX + BY + CZ + W = D,$$

$$E.*W = W. \tag{4.4}$$

Note that our formulation is more general than recent research on compressive sensing (e.g., [2, 12, 13, 17]), which do not consider anomalies, have simpler constraints (e.g., there is no $A$, $B$, or $C$), and have less general objectives.

### 4.1.2 Optimization Algorithm

The generality of the formulation in Eq. (4.4) makes it challenging to optimize. We are not aware of any existing work on compressive sensing that can cope with such a general formulation. Below we first reformulate Eq. (4.4) to make it easier to solve. We then consider the augmented Lagrangian function of the reformulated problem and develop an Alternating Direction Method to solve it.

**Reformulation for Optimization** Note that $X$ and $Y$ appear in multiple locations in the objective function and constraints in the optimization problem 4.4. This coupling makes optimization difficult. To reduce coupling, we introduce a set of auxiliary variables $X_0, X_1, \cdots, X_K$ and $Y_0$ and reformulate the problem as follows:

$$\text{minimize:} \quad \alpha \|X\|_* + \beta \|Y\|_1 + \frac{1}{2\sigma} \|Z\|_F^2$$

$$+ \frac{\gamma}{2\sigma} \sum_{k=1}^{K} \|P_k X_k Q_k^T - R_k\|_F^2,$$

$$\text{subject to:} \quad AX_0 + BY_0 + CZ + W = D,$$

$$E. * W = W,$$

$$X_k - X = 0 \quad (\forall k = 0, \cdots, K),$$

$$Y_0 - Y = 0. \tag{4.5}$$

where $Y_0$ and $X_k (0 \leq k \leq K)$ are auxiliary variables. Note that formulations Eq. (4.5) and Eq. (4.4) are equivalent.

**Alternating Direction Method for Solving** (4.5) We apply an Alternating Direction Method (ADM) [1] to solve the convex optimization problem in (4.5). Specifically, we consider the augmented Lagrangian function:

$$\mathcal{L}(X, \{X_k\}, Y, Y_0, Z, W, M, \{M_k\}, N, \mu)$$

$$\triangleq \alpha \|X\|_* + \beta \|Y\|_1 + \frac{1}{2\sigma} \|Z\|_F^2$$

$$+ \frac{\gamma}{2\sigma} \sum_{k=1}^{K} \|P_k X_k Q_k^T - R_k\|_F^2$$

$$+ \langle M, D - AX_0 - BY_0 - CZ - W \rangle \tag{4.6}$$

$$+ \sum_{k=0}^{K} \langle M_k, X_k - X \rangle \tag{4.7}$$

$$+ \langle N, Y_0 - Y \rangle \tag{4.8}$$

$$+ \frac{\mu}{2} \cdot \| D - AX_0 - BY_0 - CZ - W \|_F^2 \tag{4.9}$$

$$+ \frac{\mu}{2} \cdot \sum_{k=0}^{K} \| X_k - X \|_F^2 \tag{4.10}$$

$$+ \frac{\mu}{2} \cdot \| Y_0 - Y \|_F^2 \tag{4.11}$$

where $M$, $\{M_k\}$, $N$ are the Lagrangian multipliers [8] for the equality constraints in Eq. (4.5), and $\langle U, V \rangle \triangleq \sum_{i,j} (U[i, j] \cdot V[i, j])$ for two matrices $U$ and $V$ (of the same size). Essentially, the augmented Lagrangian function includes the original objective, three Lagrange multiplier terms (4.6)–(4.8), and three penalty terms converted from the equality constraints (4.9)–(4.11). Lagrange multipliers are commonly used to convert an optimization problem with equality constraints into an unconstrained one. Specifically, for any optimal solution that minimizes the (augmented) Lagrangian function, the partial derivatives with respect to the Lagrange multipliers must be 0. Hence the original equality constraints will be satisfied. The penalty terms enforce the constraints to be satisfied. The benefit of including Lagrange multiplier terms in addition to the penalty terms is that $\mu$ no longer needs to iteratively increase to $\infty$ to solve the original constrained problem, thereby avoiding ill-conditioning [1]. Note that we do not include terms corresponding to constraint $E. * W = W$ in the augmented Lagrangian function, because it is straightforward to enforce this constraint during each iteration of the Alternating Direction Method without the need for introducing additional Lagrange multipliers.

The ADM algorithm progresses in an iterative fashion. During each iteration, we alternate among the optimization of the augmented Lagrangian function by varying each one of $X$, $\{X_k\}$, $Y$, $Y_0$, $Z$, $W$, $M$, $\{M_k\}$, $N$ while fixing the other variables. Introducing auxiliary variables $\{X_k\}$ and $Y_0$ makes it possible to obtain a close-form solution for each optimization step. Following ADM, we increase $\mu$ by a constant factor $\rho \geq 1$ during each iteration. When involving only two components, ADM is guaranteed to converge quickly. In our general formulation, convergence is no longer guaranteed, though empirically we observe quick convergence in all our experiments (e.g., as shown in Sect. 4.2). We plan to apply techniques in [5] to ensure guaranteed convergence in future work. We further improve efficiency by replacing exact optimization with approximate optimization during each iteration. Appendix A.1 gives a detailed description on the steps during each iteration.

**Improving Efficiency Through Approximate SVD** The most time-consuming operation during each iteration of the Alternating Direction Method is performing the singular value decomposition. In our implementation, we add an additional constraint on the rank of matrix $X$: $rank(X) \leq r$, where $r$ is a user-specified parameter that represents an estimated upper bound on the true rank of $X$. We then explicitly maintain the SVD of $X$ and update it approximately during each iteration through the help of rank-revealing QR factorization of matrices that have only $r$ columns (which are much smaller than the original matrices used in SVD). We omit the details of approximate SVD in the interest of space.

### 4.1.3  Setting Parameters

**Setting $\alpha$, $\beta$ and $\sigma$**  A major advantage of our LENS decomposition is that a good choice of the parameters $\alpha$ and $\beta$ can be analytically determined without requiring any manual tuning. Specifically, let $\sigma_D$ be the standard deviation of measurement noise in data matrix $D$ (excluding the effect of low-rank, sparse, and error terms). For now, we assume that $\sigma_D$ is known, and we will describe how to determine $\sigma_D$ later in this section. Moreover, we first ignore the domain knowledge term and will adaptively set $\gamma$ for the domain knowledge term based on the given $\alpha$ and $\beta$.

Let density $\eta(D) = 1 - \frac{\sum_{i,j} E[i,j]}{m \times n}$ be the fraction of entries in $D$ that are neither missing nor erroneous, where the size of $D$ is $m \times n$ and the size of $Y$ is $m_Y \times n_Y$. $E[i, j]$ can be estimated based on domain knowledge. For example, we set $E[i, j] = 1$ if the corresponding entry takes a value outside its normal range (e.g., a negative traffic counter) or measurement software reports an error on the entry. Moreover, our evaluation shows that LENS is robust against estimation error in $\eta(D)$.

We propose to set:

$$\alpha = (\sqrt{m_X} + \sqrt{n_X}) \cdot \sqrt{\eta(D)} \tag{4.12}$$

$$\beta = \sqrt{2 \cdot \log(m_Y \cdot n_Y)} \tag{4.13}$$

$$\sigma = \theta \cdot \sigma_D \tag{4.14}$$

where $(m_X, n_X)$ is the size of $X$, $(m_Y, n_Y)$ is the size of $Y$. $\theta$ is a user-specified control parameter that limits the contamination of the dense measurement noise $\sigma_D$ when computing $X$ and $Y$. In all our experiments, we set $\theta = 10$, though it is also possible to choose $\theta$ adaptively, just like how we choose $\gamma$ as described later in this section.

Below we provide some intuition behind the above choices of $\alpha$ and $\beta$ using the basic formulation in Eq. (4.1). The basic strategy is to consider all variables except one are fixed. Our evaluation shows that these choices work well in practice.

**Intuition Behind the Choice of $\alpha$**  Consider the special case when all variables except that $X$ are already given and stay fixed. Then we just need to solve:

$$\min_{X} \quad \alpha \cdot \|X\|_* + \frac{1}{2\sigma} \cdot \|D - X - Y - W\|_F^2 \tag{4.15}$$

since $Z = D - X - Y - W$. We can prove the optimal $X$ in Eq. (4.15) can be obtained by performing soft-thresholding (a.k.a., shrinkage) on the singular values of $D - Y - W$. That is,

$$X_{\text{opt}} = \text{SVSoftThresh}(D - Y - W, \ \alpha \cdot \sigma)$$

$$\triangleq U * \text{SoftThresh}(S, \ \alpha \cdot \sigma) * V^T, \qquad (4.16)$$

where $[U, S, V] = \text{svd}(D - Y - W)$ is the singular value decomposition of $(D - Y - W)$ (thus $D - Y - W = USV^T$), and $\text{SoftThresh}(S, \ \alpha\sigma) = \text{sign}(S). * \max\{0, \ \text{abs}(S) - \alpha\sigma\}$ ($\text{sign}(S) = S./\text{abs}(S)$). Intuitively, soft-thresholding eliminates the contamination on the singular values of $X$ due to the dense measurement noise $\sigma_D$.

From asymptotic random matrix theory [10], for a random matrix with entries drawn *i.i.d.* from a Gaussian distribution with probability $\eta(D)$, its norm (i.e., the largest singular value) is bounded by $O((\sqrt{m} + \sqrt{n}) \cdot \sqrt{\eta(D)} \cdot \sigma_D)$ with a high probability. So a good heuristic is to set the soft threshold to:

$$\alpha \cdot \sigma = (\sqrt{m} + \sqrt{n}) \cdot \sqrt{\eta(D)} \cdot \sigma_D \cdot \theta,$$

where $\theta$ is a control parameter that captures the desired separation from the influence of dense measurement noise $\sigma_D$. Therefore, we simply set $\alpha = (\sqrt{m} + \sqrt{n}) \cdot \sqrt{\eta(D)}$ and $\sigma = \theta \cdot \sigma_D$.

**Intuition Behind the Choice of $\beta$**  Now suppose $X$ is given and we need to solve:

$$\min_Y \ \beta \|Y\|_1 + \frac{1}{2\sigma} \cdot \|D - X - Y - W\|_F^2 \qquad (4.17)$$

We can prove that the optimal $Y$ for (4.17) can be obtained by performing soft-thresholding (a.k.a., shrinkage) on the entries of $D - X - W$. Specifically, we have:

$$Y_{\text{opt}} = \text{SoftThresh}(D - X - W, \ \beta \cdot \sigma), \qquad (4.18)$$

where soft-thresholding eliminates the contamination on entries of $Y$ due to the dense measurement noise.

In the context of standard compressive sensing setting:

$$\min_y \ \beta * \sigma_d \cdot \|y\|_1 + \frac{1}{2} \cdot \|y - d\|_2^2,$$

where $y$ is a vector of length $n_y$, and $\sigma_d$ is the standard deviation of the vector of observables $d$. As justified in Basis Pursuit De-Noising (BPDN) [3], a penalty term $\beta = \sqrt{2 \cdot \log(n_y)}$ should be used, where $n_y$ is the number of elements in $Y$. Similarly, in our context, a good heuristic is to set the soft threshold to:

$$\beta \cdot \sigma = \sqrt{2 \cdot \log(m_Y \cdot n_Y)} \cdot \sigma_D \cdot \theta.$$

So we simply set $\beta = \sqrt{2 \cdot \log(m_Y \cdot n_Y)}$ and $\sigma = \theta \cdot \sigma_D$.

**Estimating $\sigma_D$** When $\sigma_D$ is not known in advance, we simply estimate it from entries of $D - AX_0 - BY_0 - W$ during each iteration in the Alternating Direction Method (see Appendix A.1). Specifically, let $J = D - AX_0 - BY_0 - W$, we estimate $\sigma_D$ as the standard deviation of $\{J[i, j] \mid E[i, j] = 0\}$. It is also possible to use a more robust estimator (e.g., the mean absolute value), which gives similar performance in our experiments.

**Searching for $\gamma$** $\gamma$ reflects the importance of domain knowledge terms. It is challenging to find an appropriate $\gamma$, since its value depends on how valuable are the domain knowledge versus the information from the measurement data. Therefore instead of using a fixed value, we automatically learn $\gamma$ without user feedback or ground-truth of the real missing entries as follows. Given the incomplete data matrix $D$, we further drop additional entries of $D$ and apply our algorithm under several $\gamma$ values, and quantify the error of fitting the entries that were present in $D$ but dropped intentionally during the search (so we know their true values). We adopt the value of $\gamma$ that gives the lowest fitting error on these entries as the final $\gamma$ and apply it to our final matrix interpolation, which only has the real missing elements.

**Supporting the General Formulation** In the general formulation in Eq. (4.4), we first ensure that matrices $A$, $B$, $C$ are properly scaled such that each column of $A$, $B$, $C$ has unit length (i.e. the square sum of all elements in a column is equal to 1). We also automatically scale $P_k$, $Q_k$, $R_k$ such that each row of $P_k$ and $Q_k$ has unit length. We then use the same choice of $\alpha$, $\beta$, $\sigma$, and $\gamma$ as in the basic formulation.

## 4.2 Evaluation

### 4.2.1 Evaluation Methodology

**Performance Metrics** We quantify the performance in terms of estimation error of the missing entries and anomaly detection accuracy. We drop data from existing network matrices and compare our estimation with the ground truth. We use Normalized Mean Absolute Error (NMAE) to quantify the estimation error. NMAE is defined as follow:

$$NMAE = \frac{\sum_{i,j:M(i,j)=0} |X(i, j) - \hat{X}(i, j)|}{\sum_{i,j:M(i,j)=0} |X(i, j)|}, \tag{4.19}$$

where $X$ and $\hat{X}$ are the original and estimated matrices, respectively. We only measure errors on the missing entries. For each setting, we conduct 10 random runs, which drop random set of data, and report an average of these 10 runs.

We quantify the anomaly detection accuracy using *F1-score* [15], which is the harmonic mean of precision and recall: $F1\text{-}score = \frac{2}{1/precision+1/recall}$, where precision is the fraction of anomalies found by anomaly detection schemes that are indeed real anomalies we injected and recall is the fraction of real anomalies that are correctly identified by anomaly detection schemes. The higher F1-score, the better. F1-score of 1 is perfect. We report an average of 10 random runs.

**Anomaly Generation**  As mentioned in Sect. 3.2.1, we find the maximum difference between the original trace and the EWMA prediction, and then inject the anomaly of this size to the trace. We vary the anomaly size using different scaling factors $s$ and the fraction of anomalies to understand their impacts.

**Different Dropping Modes**  As in [17], we drop data in the following ways: (i) PureRandLoss: elements in a matrix are dropped independently with a random loss rate; (ii) xxTimeRandLoss: xx% of columns in a matrix are selected and the elements in these selected columns are dropped with a probability $p$ to emulate random losses during certain times (e.g., disk becomes full); (iii) xxElemRandLoss: xx% of rows in a matrix are selected and the elements in these selected rows are dropped with a probability $p$ to emulate certain nodes lose data (e.g., due to battery drain); (iv) xxElemSyncLoss: the intersection of xx% of rows and p% of columns in a matrix are dropped to emulate a group of nodes experience the same loss events at the same time; (v) RowRandLoss: drop random rows to emulate node failures, and (vi) ColRandLoss: drop random columns for completeness. We use PureRandLoss as the default, and further use other loss models to understand impacts of different loss models. We feed the matrices after dropping as the input to LENS, and use LENS to fill in the missing entries.

**Schemes Evaluated**  We compare the following schemes:

- Base: It approximates the original matrix as a rank-2 approximation matrix $X_{base} = \overline{X} + X_{row}1^T + 1X_{col}^T$, where 1 is a column vector consisting of all ones and $X_{row}$ and $X_{col}$ are computed using a regularized least square according to [17].
- SVD Base: As shown in [17], SVD Base, which applies SVD to $X - X_{base}$, out-performs SVD applied directly to $X$. We observe similar results, so below we only include SVD Base.
- SVD Base + KNN: We obtain the result from SVD Base and then apply $K$ nearest neighbors (KNN) to perform local interpolation to leverage the local structure.
- SRMF: Sparsity Regularized Matrix Factorization (SRMF) leverages both low-rank and spatio-temporal characteristics [17].
- SRMF+KNN: It combines SRMF results with local interpolation via KNN [17].
- LENS: We use the output from the LEN decomposition as described in Sect. 4.1.

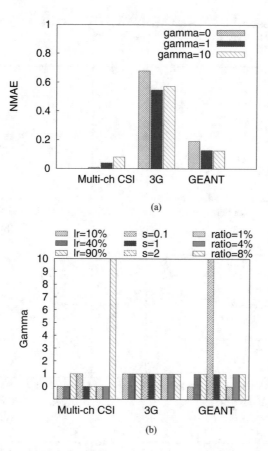

**Fig. 4.1** Self learned $\gamma$. (**a**) NMAE of different $\gamma$. (**b**) $\gamma$ under different conditions

## 4.2.2 Performance Results

**Self Learned** $\gamma$  LENS supports many types of domain knowledge as described in Sect. 4.1.1. Our evaluation only considered temporal stability for simplicity and $\gamma$ reflects its importance. To illustrate the benefit of self learning, Fig. 4.1a shows the performance under different $\gamma$ values and different traces. Figure 4.1b shows the best $\gamma$ under different traces, loss rates, anomaly sizes, and ratio of anomalies. There does not exist a single $\gamma$ that works well for all traces or conditions. Self tuning allows us to automatically select the best $\gamma$ for these traces and achieves low NMAE in all cases.

**Varying Dropping Rates**  We first compare different schemes in terms of interpolation accuracy measured using NMAE. Figures 4.2 and 4.3 show the interpolation error as we randomly inject anomalies to 5% elements with $s = 1$. For clarity of the graphs, we cap the y-axis so that we can focus on the most interesting parts of the graphs. We observe that LENS consistently out-performs the other schemes.

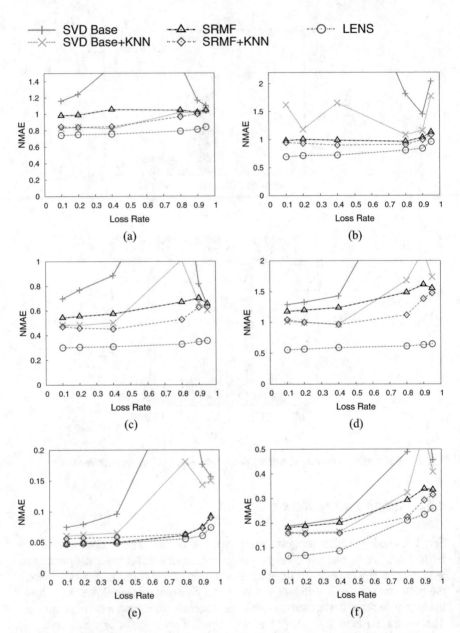

**Fig. 4.2** Interpolation performance under varying data loss rates (part I). (**a**) 3G. (**b**) WiFi. (**c**) Abilene. (**d**) GÉANT. (**e**) 1-channel CSI. (**f**) Multi-channel CSI

In terms of NMAE, LENS < SRMF + KNN < SRMF < SVD Base + KNN < SVD Base. LENS reduces NMAE by 35.5% over SRMF, 27.8% over SRMF+KNN, 59.8% over SVD Base, and 44.9% over SVD Base + KNN on average. Moreover, the error is low for high-rank matrices. For example, the highest rank matrices in our

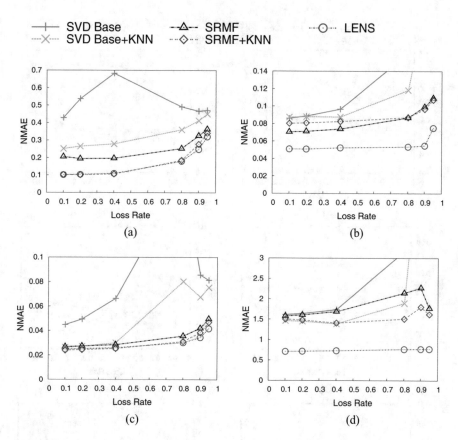

**Fig. 4.3** Interpolation performance under varying data loss rates (part II). (**a**) Cister RSSI. (**b**) CU RSSI. (**c**) UMich RSS. (**d**) UCSB Meshnet

datasets are 1-channel CSI, CU RSSI, Abilene, WiFi, 3G, and Cister RSSI matrices. Their corresponding NMAE are 0.05, 0.05, 0.3, 0.69, 0.74, 0.1, respectively. Most of them have low errors except WiFi and 3G. The error does not monotonically increase with the loss rate because an increasing loss rate reduces the number of anomalies, which may help reduce the error.

Figures 4.4 and 4.5 summarize the results under varying data loss rates and no anomaly. In most traces, LENS performs comparably to SRMF+KNN, the best known algorithm under no anomaly. In UCSB Meshnet, LENS already out-performs SRMF+KNN even without injecting additional anomalies. This is likely because the trace has more anomalies before our anomaly injection. In UCSB Meshnet trace, 3.2% of EWMA prediction errors are larger than 5 times standard deviation from mean, whereas the corresponding numbers in other traces are 1.2%–2.4%. 3G trace has the second largest number of EWMA prediction error where we can also see LENS shows 7.7% improvement over SRMF+KNN.

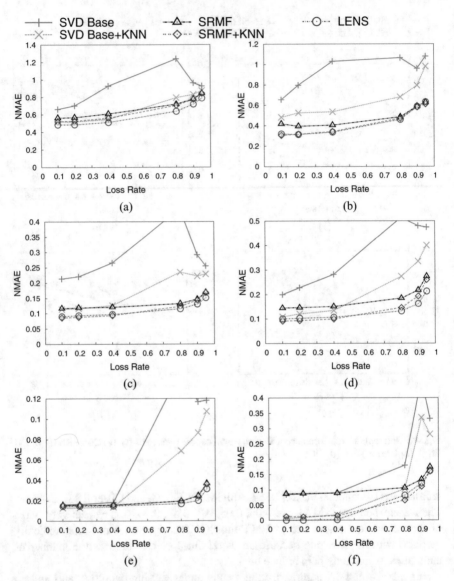

**Fig. 4.4** Varying data loss rates and no anomaly (part I). (**a**) 3G. (**b**) WiFi. (**c**) Abilene. (**d**) GÉANT. (**e**) 1-channel CSI. (**f**) Multi-channel CSI

**Varying Anomaly Sizes** Figures 4.6 and 4.7 show the interpolation performance as we vary the anomaly size $s$. LENS significantly out-performs all the other schemes. Its benefit increases with the anomaly size. For example, when $s = 1$, the NMAE of LENS is 33.7% lower than SRMF, 20.2% lower than SRMF+KNN, 61.8% lower than SVD Base, and 34.8% lower than SVD Base+KNN. The corresponding numbers under $s = 2$ are 44.9%, 31.9%, 69.8%, and 45.8%, respectively. Moreover,

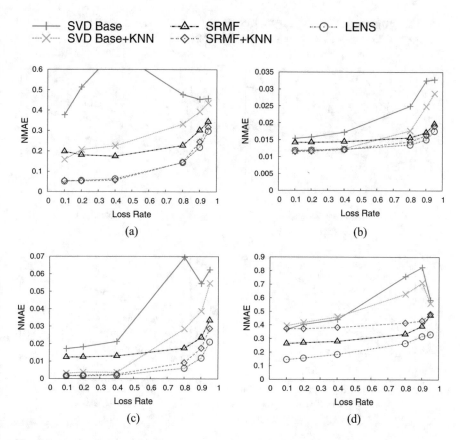

**Fig. 4.5** Varying data loss rates and no anomaly (part II). (**a**) Cister RSSI. (**b**) CU RSSI. (**c**) UMich RSS. (**d**) UCSB Meshnet

as we would expect, NMAE of all schemes tends to increase with the anomaly size in all traces, though the NMAE of LENS increases more slowly than the other schemes, since LENS explicitly separates anomalies before data interpolation. These results highlight the importance of anomaly detection in interpolation.

**Varying the Number of Anomalies** Figures 4.8 and 4.9 show the interpolation performance as we vary the number of anomalies. As before, LENS out-performs SRMF and SVD based schemes. The improvement ranges between 25.3–59.7% with 8% anomalies and 30.1–54.5% with 16% anomalies. In addition, the NMAE increases with the number of anomalies. Among different schemes, the rate of increase is slowest in LENS due to its explicit anomaly detection and removal.

**Varying Noise Sizes** Figures 4.10 and 4.11 show the interpolation performance as we vary the noise sizes. We inject noise to all the elements in the original matrices. The size of the noise follows normal distribution with mean 0 and standard deviation

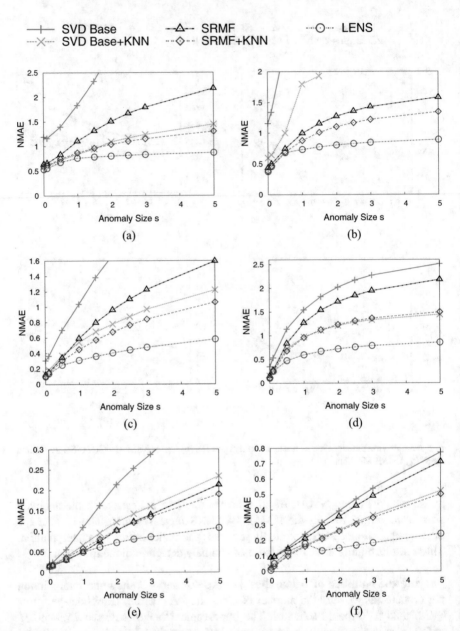

**Fig. 4.6** Impact of anomaly sizes (part I). (**a**) 3G. (**b**) WiFi. (**c**) Abilene. (**d**) GÉANT. (**e**) 1-channel CSI. (**f**) Multi-channel CSI

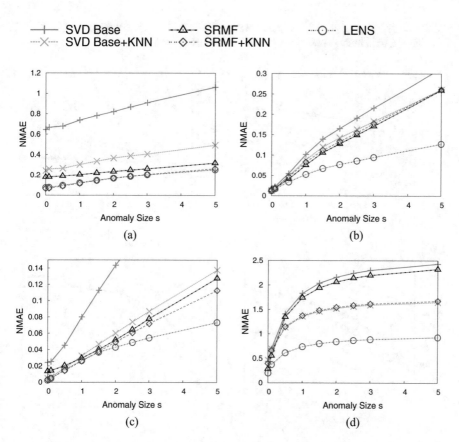

**Fig. 4.7** Impact of anomaly sizes (part II). (**a**) Cister RSSI. (**b**) CU RSSI. (**c**) UMich RSS. (**d**) UCSB Meshnet

$\sigma$ where $\sigma$ is varied from 1% to 64% of the maximal value in the matrix. As before, LENS out-performs the other schemes.

**Different Dropping Modes** Next we compare the interpolation accuracy under different dropping modes. In the interest of brevity, Fig. 4.12 shows interpolation error for UCSB Meshnet traces. NMAE is similar for the other traces. As we can see, LENS yields lowest NMAE under all dropping modes. It out-performs SRMF-based schemes by 52.9%, and out-perform SVD-based schemes by 60.0%.

**Prediction** Prediction is different from general interpolation because consecutive columns are missing. SVD is not applicable in this context. KNN does not work well either since temporally or spatially near neighbors have missing values. Figure 4.13 shows the prediction error as we vary the prediction length (i.e., prediction length $l$ means that the first $1 - l$ columns are used to predict the remaining $l$ columns). We include Base in the figure since [17] shows Base is effective in prediction. LENS out-performs SRMF, which out-performs Base.

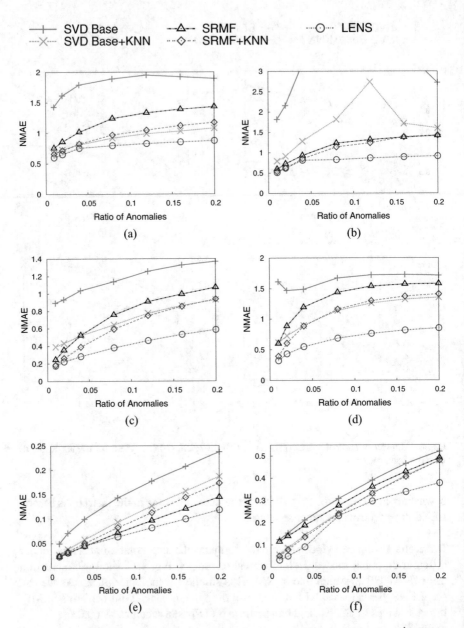

**Fig. 4.8** Impact of number of anomalies (part I). (**a**) 3G. (**b**) WiFi. (**c**) Abilene. (**d**) GÉANT. (**e**) 1-channel CSI. (**f**) Multi-channel CSI

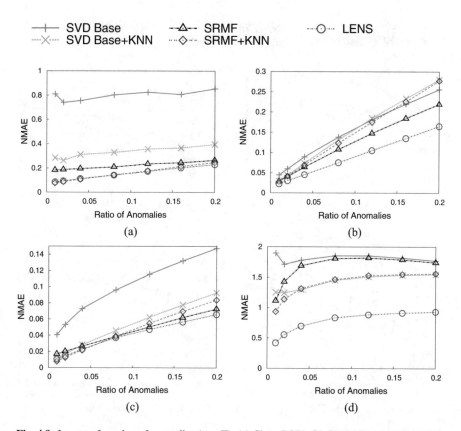

**Fig. 4.9** Impact of number of anomalies (part II). (**a**) Cister RSSI. (**b**) CU RSSI. (**c**) UMich RSS. (**d**) UCSB Meshnet

Figure 4.14 further compares Base, SRMF, and LENS as we vary anomaly size. LENS continues to out-perform SRMF and Base. On average, it improves SRMF by 17.7%, and improves Base by 30.4%. Figure 4.15 shows the performance as we vary the number of anomalies. LENS continues to perform the best, out-performing SRMF by 29.6% and Base by 34.6%.

**Anomaly Detection** We further compare the accuracy of anomaly detection as we inject anomalies to 5% elements with $s = 1$. SRMF detects anomalies based on the difference between the actual and estimated values, and consider the entry has an anomaly if its difference is larger than a threshold. LENS considers all entries whose Y values are larger than a threshold as anomalies. Following [17], for each of the schemes, we choose a threshold to achieve the false alarm probability within $10^{-5}$. As shown in Fig. 4.16, LENS consistently out-performs SRMF+KNN. In 3G and Cister RSSI traces, its F1-score is 17.6% higher than that of SRMF+KNN. This shows that LENS is effective in anomaly detection.

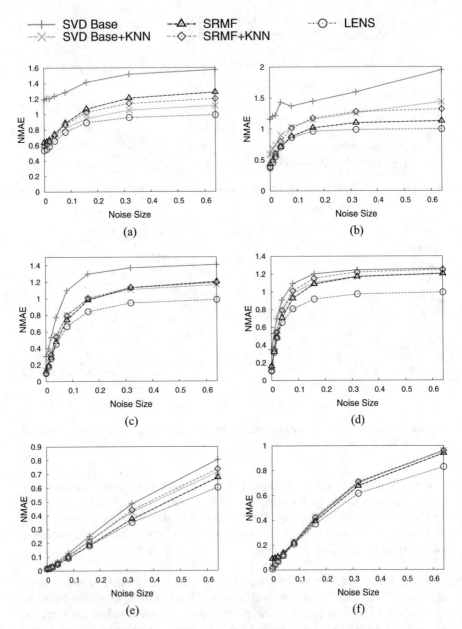

**Fig. 4.10** Impact of noise sizes (part I). (**a**) 3G. (**b**) WiFi. (**c**) Abilene. (**d**) GÉANT. (**e**) 1-channel CSI. (**f**) Multi-channel CSI

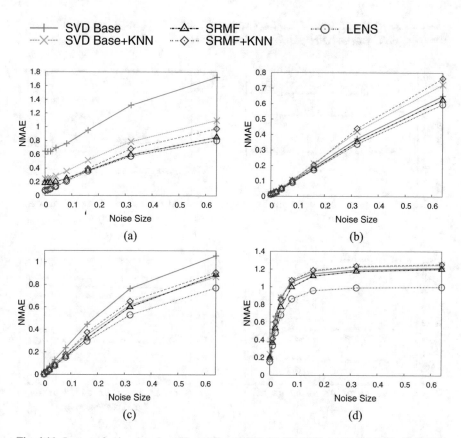

**Fig. 4.11** Impact of noise sizes (part II). (**a**) Cister RSSI. (**b**) CU RSSI. (**c**) UMich RSS. (**d**) UCSB Meshnet

**Computational Time** Figure 4.17 compares the computation time of LENS and SRMF when both use 500 iterations. As we can see, LENS has much smaller computation time due to local optimization in ADM. This makes it feasible to perform efficient search over different parameters. Figure 4.18 further shows that LENS converges within 200–250 iterations.

## 4.3 Conclusion

This chapter presents *LENS decomposition* to decompose a network matrix into a low-rank matrix, a sparse anomaly matrix, an error matrix, and a dense but small noise matrix. Our evaluation shows that it can effectively perform missing value interpolation, prediction, and anomaly detection and out-perform state-of-the-art approaches. As part of our future work, we plan to apply our framework to network

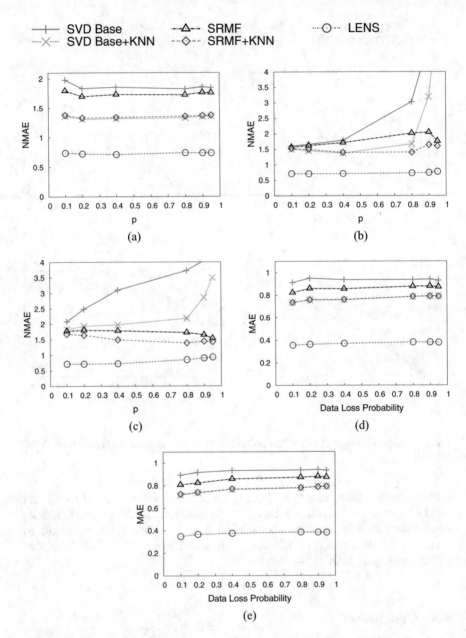

**Fig. 4.12** UCSB Meshnet: varying dropping models. (**a**) 50TimeRandLoss. (**b**) 50ElemRandLoss. (**c**) 50ElemSyncLoss. (**d**) RowRandLoss. (**e**) ColRandLoss

tomography (e.g., traffic matrix estimation based on link loads and link performance estimation based on end-to-end performance). As part of our future work, we plan to apply LENS to enable several important wireless applications, including spectrum sensing, channel estimation, and localization.

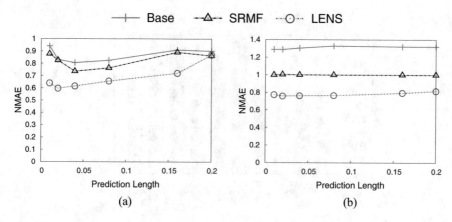

**Fig. 4.13** Prediction performance under 5% anomalies. (**a**) WiFi. (**b**) UCSB Meshnet

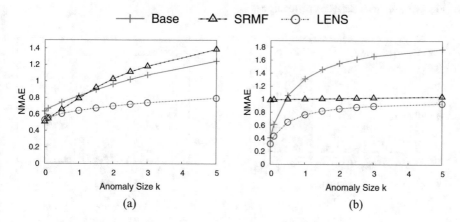

**Fig. 4.14** Prediction performance with various anomaly sizes. (**a**) WiFi. (**b**) UCSB Meshnet

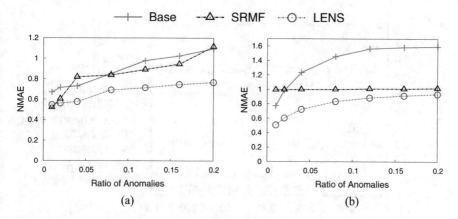

**Fig. 4.15** Prediction performance with various number of anomalies. (**a**) WiFi. (**b**) UCSB Meshnet

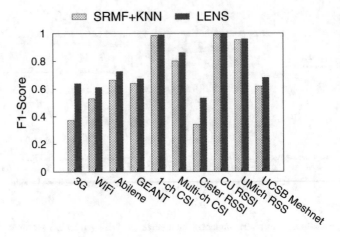

**Fig. 4.16** Anomaly detection performance

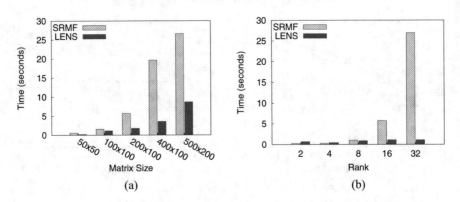

**Fig. 4.17** Computation time. (**a**) vary matrix sizes (rank=10). (**b**) vary ranks ($100 \times 100$ matrices)

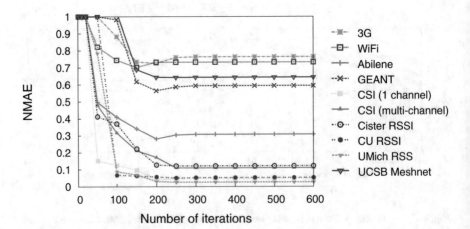

**Fig. 4.18** The performance under various number of iterations

# References

1. Augmented lagrangian method (2022). http://en.wikipedia.org/wiki/Augmented_Lagrangian_method
2. E.J. Candes, X. Li, Y. Ma, J. Wright, Robust principal component analysis, 2009. Manuscript. Available from http://www-stat.stanford.edu/~candes/papers/RobustPCA.pdf
3. S.S. Chen, D.L. Donoho, M.A. Saunders, Atomic decomposition by basis pursuit. SIAM J. Sci. Comput. **20**, 33–61 (1998)
4. G. Cheng, H. Chen, D. Cheng, Z. Zhang, J. Lan, Anomaly detections in internet traffic using empirical measures. Int. J. Innov. Technol. Explor. Eng. (2013). https://doi.org/10.1145/1452520.1452539
5. W. Deng, M.-J. Lai, Z. Peng, W. Yin, Parallel multi-block ADMM with o(1/k) convergence, in *UCLA CAM 13–64*, 2014
6. B. George, S. Kim, S. Shekhar, Spatio-temporal network databases and routing algorithms: a summary of results, in *Proc. of SSTD* (Springer, Berlin, Heidelberg, 2007), pp. 460–477
7. J. Kowalski, B. Warfield, Modeling traffic demand between nodes in a telecommunications network, in *Proc. of ATNAC*, 1995
8. Lagrange multiplier (2022). http://en.wikipedia.org/wiki/Lagrange_multiplier
9. L.F. Lu, Z.-H. Huang, M.A. Ambusaidi, K.-X. Gou, A large-scale network data analysis via sparse and low rank reconstruction, in *Discrete Dynamics in Nature and Society*, February 2013
10. D. Needell, Non-asymptotic theory of random matrices. Lecture 6: norm of a random matrix, 2007. Stanford Math 280 Lecture Notes. Available at http://www-stat.stanford.edu/~dneedell/lecs/lec6.pdf
11. S. Rallapalli, L. Qiu, Y. Zhang, Y.-C. Chen, Exploiting temporal stability and low-rank structure for localization in mobile networks, in *Proc. of ACM MobiCom*, 2010
12. B. Recht, M. Fazel, P.A. Parrilo, Guaranteed minimum-rank solutions of linear matrix equations via nuclear norm minimization, in *SIAM Review*, 2007
13. B. Recht, W. Xu, B. Hassibi, Necessary and sufficient condtions for success of the nuclear norm heuristic for rank minimization, in *Proc. of Decision and Control*, July 2008
14. M. Wang, A. Ailamaki, C. Faloutsos, Capturing the spatio-temporal behavior of real traffic data. Perform. Eval. **49**(1–4), 147–163 (2002)
15. Wikipedia, F1 score (2022). http://en.wikipedia.org/wiki/F1_score
16. Y. Zhang, M. Roughan, N. Duffield, A. Greenberg, Fast accurate computation of large-scale IP traffic matrices from link loads, in *Proc. ACM SIGMETRICS*, June 2003
17. Y. Zhang, M. Roughan, W. Willinger, L. Qiu, Spatio-temporal compressive sensing and internet traffic matrices. ACM SIGCOMM Comput. Commun. Rev. **39**(4), 267 (2009)

# Chapter 5
# Data-Driven Synchronization

In this chapter, we present the data-driven synchronization algorithm and apply it to improve interpolation, enable segmentation, and perform activity recognition.

## 5.1 Motivation

The lack of synchronization and uniform sampling rates lead to network-induced blurring, and can easily cause a low-rank matrix to become a much higher rank. As an example, Fig. 5.1a shows that 8 identical signals with different shifts become less correlated and the matrix formed by these signals is full rank due to misalignment. Figure 5.1b shows that 8 identical signals with different frequencies appear rather distinct and the matrix formed by these signals is full rank.

Ranks of real-world data analyzed in Sect. 3.2.2 support our observation. It highlights the importance of synchronization. Synchronizing the traces reduces the ranks and increases effectiveness of compressive sensing, which assumes low-rank matrices. Moreover, it increases the spatial locality in the data, which helps segmentation and activity recognition.

## 5.2 Data-Driven Synchronization

Our goal is to synchronize time series from different users or nodes to maximize spatial locality, which helps improve interpolation, segmentation, and activity recognition. Synchronization needs to account for the time and frequency-domain heterogeneity of the real-world data and therefore depends on the characteristics of traces. For periodic traces, we should first bring all the traces to the uniform frequencies and then synchronize the alignment between them. As an example,

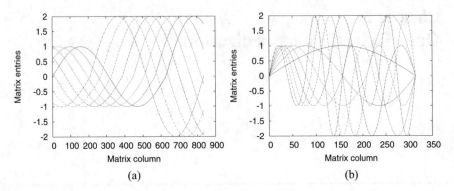

**Fig. 5.1** Effects of different offsets and frequencies. (**a**) Effect of misalignment. (**b**) Effect of different frequencies

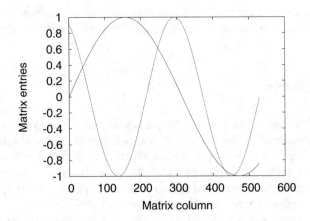

**Fig. 5.2** Synchronization in periodic traces

consider two time series shown in Fig. 5.2. The two time series not only have misalignment, but also have different frequencies. We need to first convert them to the same frequency before finding the shift that aligns them. On the other hand, for non-periodic traces, it is sufficient to synchronize them directly. Therefore, our approach first separates periodic from non-periodic traces, and then synchronizes them accordingly.

**Identifying Periodic Portions** We use a sliding window to determine if the current window contains periodic traces. We compute normalized auto-correlation with a varying lag $\tau$ as in [10]. Specifically, we calculate the normalized auto-correlation for lag $\tau$ at time $t$:

$$\rho(t, \tau) = \frac{E[(X_t - \mu_t)(X_{t+\tau} - \mu_{t+\tau})]}{\tau \sigma_t \sigma_{t+\tau}} \tag{5.1}$$

where $\mu_t$ is the mean of the time series in the current window $[t, t+\tau-1]$, $\sigma_t$ is the standard deviation, and $E[\cdot]$ is the expected value operator. A signal with a period $\tau$ has $\rho(t, \tau)$ close to 1. To avoid treating a constant time series (which also has a high $\rho$) as periodic, we require the standard deviation of the signal in the current window is higher than a threshold (e.g., 0.1 used in our evaluation) and there is a lag $\tau$ that gives a jump in auto-correlation for the data to be considered as periodic. The lag $\tau$ is not known in advance. In our traces, the periodic activities are due to walking and climbing stairs. Given the sampling rates of 32-52Hz, the typical 2-step duration should be 18-150 samples, so we search the lag $\tau$ in this window.

**Synchronizing Non-periodic Traces** The traces are likely to exhibit spatial locality in the measurements from different nodes (i.e., they may be grouped into a small number of clusters). Unlike traditional clustering, which clusters raw data directly, we use synchronization to enhance the spatial locality and improve clustering. A major challenge arises from the strong interactions between synchronization and clustering (i.e., which nodes to synchronize affects which nodes from a cluster, and vice versa).

To address this challenge, we develop a novel iterative clustering algorithm that simultaneously synchronizes and clusters nodes. We first randomly select $K$ rows as the cluster heads. For each of the remaining rows in the matrix, we search for the offset that maximizes the cross correlation coefficient with each of the $K$ cluster heads and assign it to the cluster that has the highest correlation coefficient. After all the remaining rows have been assigned to the cluster, we select the new cluster head that has the highest correlation coefficient with all the other cluster member (i.e., $i$ such that $max_i \sum_{j \neq i} \rho_{i,j}$, where $i$ and $j$ belong to the same cluster). After selecting new cluster heads, we iterate by assigning the remaining rows to the cluster head that has the highest correlation coefficient. The iteration continues until the average increase in the correlation coefficient across clusters is smaller than a threshold or the maximum number of iterations has reached. Our evaluation uses 0.001 as the threshold for improvement in correlation coefficient, and 1000 as the maximum iterations. Usually clustering converges within 10 iterations.

Several comments follow. First, a well known challenge in clustering algorithm is how to select the number of clusters $K$. We address this by automatically searching for $K$ that yields the best performance. To quantify the performance, we apply synchronization to interpolation, drop additional elements in the matrix, and use the $K$ that gives the lowest interpolation error on the intentionally dropped entries (since we know their ground-truth values). Second, we try various metrics for synchronization, including correlation coefficient, DTW distance, and rank, and find synchronization based on correlation coefficient yields highest interpolation accuracy. Third, some traces have multiple features associated with each user. In such a case, which feature should we use for synchronization? We examine two options: the average correlation coefficients across all features and the correlation coefficient using the best feature (i.e., the feature that has the highest correlation coefficient.) Our evaluation shows the best feature yields higher synchronization

```
1  Function SyncNonPeriod (traces, K)
2  |    heads = Rand (K) ; // randomly select K heads
3  |    for iter = 1 to maxIter do
4  |    |    clusters = [heads] // create clusters with the selected heads
5  |    |    for i = 1 to N do
6  |    |    |    for c in clusters do
7  |    |    |    |    find offset s.t. maximizes corrcoef(i, heads(c));
8  |    |    |    end
9  |    |    |    add i to the cluster c that has max corrcoef(i, heads(c))
10 |    |    end
11 |    |    for c in clusters do
12 |    |    |    for i in c do
13 |    |    |    |    avg_corrcoef(i) = mean_{i≠j,i,j∈c}corrcoef(i, j);
14 |    |    |    |    update heads(c) to i if avg_corrcoef(i) is higher
15 |    |    |    end
16 |    |    end
17 |    |    if Δ∑_c avg_corrcoef(heads(c)) < thresh then
18 |    |    |    break;
19 |    |    end
20 |    end
21 |    return clusters;
22 end
```

**Algorithm 1:** Synchronization for non-periodic traces

accuracy. Finally, we should emphasize that synchronization should be performed at the cluster level (i.e., all members in the cluster synchronize to the same node) instead of pairwise since all members should synchronize with each other to reduce the rank of the entire matrix and improve interpolation accuracy.

**Synchronizing Periodic Traces** The above algorithm works well for non-periodic data. For periodic data with different frequencies, we first bring them to the same frequencies before searching for the offset to synchronize. We run Short-Time Fourier Transform (STFT) to determine the dominant frequency in each row. We choose one row as the target, and uniformly stretch each of the remaining rows based on the ratio between the dominant frequency of that row and target row. Stretching is achieved by adding points. We support a non-integer stretching ratio by finding a least common multiple of the dominant frequencies in the two time series and stretch each trace towards the least common multiple. For example, the two traces have frequencies of 4 and 6, respectively. We stretch trace 1 to a frequency of 12 by adding two samples for every real sample, and stretch trace 2 to the same frequency by adding one sample for every real sample. After bringing the rows to the same dominant frequency, we search for the offset to maximize the correlation coefficient. The above algorithm assumes each row has one dominant frequency. If a row has multiple dominant frequencies (e.g., an user walks at different speeds over time), we partition the time series into two and find dominant frequency in each, and recurse if needed. Once we find the dominant frequency for each portion, we convert all portions into the same frequency using the above algorithm.

```
1  Function SyncPeriod (traces, K)
2      for t in traces do
3          freq = FindDomFreq (t) ; //calculate dominant freq
4          //uniformly stretch a trace based on its dominant freq
5          post_traces(i++) = StretchTraces (t, freq) ;
6      end
7      clusters = SyncNonPeriod (post_traces, K) ;
8      return clusters;
9  end
```

**Algorithm 2:** Synchronization for periodic traces

With the above modification in mind, we have the following clustering algorithm. As before, we randomly select $K$ rows as cluster heads. For each of the remaining rows, we try to synchronize it to each of the $K$ cluster heads by first synchronizing their frequencies and then finding the offset that maximizes the correlation coefficient. We then assign the row to the cluster head that yields the largest correlation coefficient. After all rows have been assigned, we select new cluster heads as the ones having the largest correlation coefficient with all cluster members, and iterate until the improvement is within a threshold or the maximum number of iterations is reached.

## 5.3 Applications

Next we show how to apply data-driven synchronization to improve interpolation, segmentation, and activity recognition.

### 5.3.1 Interpolation

Compressive sensing is the state-of-the-art technique for interpolation. Real-world traces without synchronization may violate the low rank assumption in existing compressive sensing techniques, and significantly degrade the interpolation accuracy. We apply synchronization to enhance the effectiveness of compressive sensing algorithms.

The major challenge to support interpolation with our synchronization method is that synchronization shifts rows in a matrix by different offsets and the original matrix after the shift is no longer a matrix, as shown in Fig. 5.3a. In order to apply compressive sensing, we have to convert it back to a regular matrix. One possibility is to draw a bounding box around all rows and treat the additional elements in the bounding box as missing elements. However, this introduces many new missing elements and reduces the interpolation accuracy. Alternatively, we can use only the overlapping portion to create the matrix and perform interpolation. This is

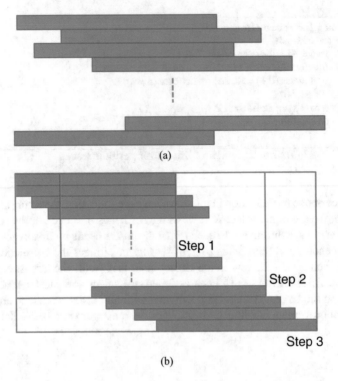

(a)

Step 1

Step 2

Step 3

(b)

**Fig. 5.3** (**a**) shows an example of matrix after synchronization, and (**b**) illustrates the interpolation algorithm

insufficient because not all missing values in the original matrix are included, and the overlapping portion may be too small for us to effectively exploit redundancy and structure in the matrix, thereby degrading interpolation accuracy.

Ideally, we want to have as a large matrix as possible without adding many missing values to achieve high interpolation accuracy. We prefer a large matrix since the low-rank property and redundancy is more evident in a larger matrix. Based on this observation, we develop an interpolation scheme that can be applied to any compressive sensing algorithm.

In the first step, we search for the matrix that has the largest overlapping area. The largest one may not include all rows, since including all rows may lead to fewer overlapping columns as only the columns overlapped by all the rows are included in the first step. So we need to select an appropriate number of rows to include to maximize the matrix size. The search space is exponential since any row may or may not be included in the largest overlapping area. We develop the following heuristic to search more efficiently. For each row $i$, we calculate the number of overlapping columns it has with another row $j$ (denoted as $OverlapColumn(i, j)$), and sort $j$ in the decreasing order of $OverlapColumn(i, j)$. We then consider the matrices that include only 1 row (the $i$-th row), 2 rows (the $i$-th row and the other row that has maximum overlap with $i$), up to $N$ rows, where $N$ is the number of rows in the original matrix. Let $M(i)$ denote the largest one among these $N$ matrices associated with row $i$. Similarly, we compute $M$ for each of the other rows. Then we pick the matrix $j$ that has the maximum size (i.e., $max_j|M(j)|$ where $|\cdot|$ denotes the matrix size), and interpolate for this matrix.

```
1  Function FindMaxOverlapArea (traces)
2  |   for each row i do
3  |   |   for each row j and j ≠ i do
4  |   |   |   OverlapColumn(i,j) ← number of overlap columns of row i and j
5  |   |   end
6  |   |   sorted_tr = Sort (OverlapColumn(i,:)) ;
7  |   |   for j = 1 to N do
8  |   |   |   mat_row(j) leftarrow find overlap matrix in sorted_tr(1:j)
9  |   |   end
10 |   |   mat(i) = Max (mat_row) ;
11 |   end
12 |   matrix = Max (mat) ;
13 |   return matrix;
14 end
```

**Algorithm 3:** Interpolation step one: find the matrix with the maximal overlapping area

Next we expand the matrix to include additional missing elements by treating the interpolated values in the previous step as known and use compressive sensing to interpolate the remaining missing elements. This process continues until we interpolate all missing values in the original matrix. To balance interpolation accuracy and running time, we use three iterations in our evaluation (i.e., we run compressive sensing three times, one on the small matrix found in the first step, one on the medium matrix including half of the remaining rows after the first step, and one on the matrix including all rows).

Figure 5.3b shows an example. We first interpolate the elements in the innermost bounding box. The missing values in this matrix are likely to have lower error since we do not add any missing elements. Then we expand the matrix to the middle bounding box, and interpolate for additional missing elements. Finally, we expand the matrix to the outermost bounding box, which includes all rows, and interpolate the remaining missing values.

## 5.3.2 Segmentation

Next we examine the segmentation (e.g., partition a timeseries into smaller pieces where each contains one activity). Some of the existing works [2, 6, 9, 11] segment the traces involving multiple activities by first identifying the type of activity each user is engaged in and then finding the transition point between two activities.

Its accuracy is limited by the accuracy of activity recognition. Moreover, it only considers one user at a time, which further limits the accuracy.

If all users perform the same activities over time (e.g., they watch the same screen with different images over time as in P300 traces, watch the same movie, listen to the same radio station, walking or climbing together as in the periodic portions in most of our activity traces), then we can leverage the time series collected from different users to enhance the accuracy. This is possible due to data-driven synchronization. We use a sliding window and compute the average correlation coefficients across all users and all features in the current sliding window. If the sliding window spans multiple activities, the correlation coefficient is high since the patterns in different activities are rather different and different users experience synchronized changes in the activities. On the other hand, when the sliding window moves over to include only one activity, there is no drastic synchronized change, which reduces correlation coefficients. So we search for the segment boundary as the one with the minimum correlation.

Figure 5.4 shows an example, where the top figure shows raw EMG traces across different users and the red lines mark the real boundaries between different activities. We see the average correlation coefficient over all users and all features (in the bottom figure) drops around the red boundaries in the top figure. So we use this heuristic to segment traces into smaller portions, each containing one activity. Unlike synchronization, segmenting using all features out-performs segmenting using the best feature. This is because segmentation requires us to support different activities, and different features work well for different activities whereas synchronizing one of the activities may allow us to synchronize the entire trace.

Our implementation first smooths the time series of the correlation coefficient using the robust spline smoothing algorithm [4] and finds the local minimums [5]. Near the boundary between the activities, correlation coefficients may fluctuate (e.g., due to random motions between activities). The fluctuation results in multiple

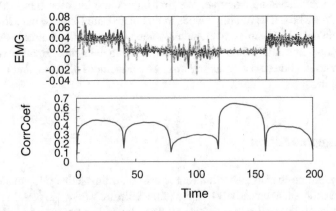

**Fig. 5.4** The average correlation coefficient over a sliding window drops around the boundaries between different activities in the EMG traces, where the sliding window size is set to the shortest activity duration

local minimums near the boundary. Therefore, when there are multiple local minimums next to each other (i.e., the distance between local minimums is smaller than a threshold, set to 20 samples in our implementation), we pick the minimal one and filter the others.

### 5.3.3 Activity Recognition

There are many algorithms for activity recognition. Existing activity recognition schemes are based on traces from individual users. When users perform the same activities, we can leverage the group information in the traces from multiple users (or runs) simultaneously to enhance the accuracy of recognition. Our group based activity recognition can apply to different recognition algorithms. Here we demonstrate the idea using linear regression. We organize the training data for each user into a matrix where different rows represent different instances of runs (or users) and different columns represent different features collected from all sensors when performing an activity. We learn the weight of each feature based on the labeled data. For activity $i$, we assign $b_i = 1$ if the current measurements are from the activity $i$, and $b_i = 0$ otherwise. Let $x_i$ denote the weight of each feature associated with activity $i$. We learn $x_i$ by solving the following optimization problem:

$$\min_{\mathbf{x}_i} \|b_i - A\mathbf{x}_i\|_2^2 + \lambda \|\mathbf{x}_i\|_2^2 \tag{5.2}$$

where $\| \cdot \|_2$ denotes the $L_2$ norm, $\lambda$ is a constant that balances the importance of minimizing the fitting error versus minimizing the magnitude of $x_i$. This problem can be efficiently solved using a least-square solver. The intuition is that we want to minimize the fitting error as much as possible, but we allow some deviation due to measurement errors. Meanwhile, we prefer a parsimonious explanation (i.e., a solution with small $L_2$ norm due to a small number of important features).

For each activity, we perform the above linear regression to learn the weight of features associated with that activity. Then we apply the weights to the testing data by computing $A_{test}\mathbf{x}_i$ for each activity $i$. We assign the current measurement to the activity $i$ that has the highest value in $A_{test}\mathbf{x}_i$.

Next we propose taking advantage of group information. Suppose all the users are doing the same group activities. We can perform voting based on the estimations from all users. A natural approach is to perform majority voting, which treats all users' estimation equally. To further improve it, we use weighted average of users' estimation, where the weight is $\frac{1}{var}$ and $var$ is the variance of the estimation. This is inspired by the maximum ratio combining (MRC) in wireless communication. The higher the variance, the less weight the estimation carries. To compute the variance, we allocate a large portion of training data $TD$ to learn the weights and use the remaining small portion of training data $TD'$ for testing. Since we know the ground truth for all training data (including $TD'$), we can compute the prediction accuracy $p$ using $TD'$. The prediction process tends to follow Bernoulli distribution, and the variance of Bernoulli distribution is $p(1 - p)$. Therefore, we can combine different users' estimation as $max_i \sum_u b_{u,i}/(p_u(1 - p_u))$, where $b_{u,i} = 1$ if the user $u$'s estimated activity is $i$, and $b_{u,i} = 0$ otherwise, and assign the current testing data to the activity $i$ with the largest weighted sum.

## 5.4   Evaluation

We evaluate the accuracy of synchronization, interpolation, segmentation, and activity recognition in turn.

### 5.4.1   Evaluation of Synchronization

Our evaluation consists of differentiating between periodic and non-periodic patterns, and synchronizing each of them.

**Accuracy of Separating Periodic Patterns**   We visually check the time series, and label samples as periodic and non-periodic patterns. The labels are used as ground truth values. We then use the algorithm described in Sect. 5.2 to separate periodic and non-periodic patterns.

Figure 5.5 shows the time series of acceleration and auto-correlation of three accelerometer traces. The period with high auto-correlation and variation are automatically identified as periodic patterns. The start and end of the identified periods are marked by vertical red lines in the top accelerometer traces. As we can see, the accuracy is quite high: most periodic portions of the traces are correctly marked as periodic.

Figure 5.6 quantifies the accuracy of separating periodic and non-periodic patterns. Precision denotes the fraction of samples that are identified as periodic are actually periodic. Recall represents the fraction of samples that are periodic are identified correctly as periodic. F1-score is the harmonic mean of precision and recall (i.e., $\frac{2}{1/precision+1/recall}$). These three metrics are $0.82 - 0.99$, indicating high accuracy.

**Synchronization Accuracy for Non-periodic Traces**   We choose the traces that were manually synchronized by the authors for evaluation so that we know the ground truth. Figure 5.7 shows the CDF of synchronization error for non-periodic traces. As we can see, all subjects in EMG, Acc-PAMAP2, and Acc-Wrist are synchronized well: normalized error (i.e., the ratio of the offset to the trace length) $< 2\%$. Acc-Chest is slightly worse where 10% of subjects have 10%–12% error. The synchronization error of P300 EEG is larger since the correlation coefficient between different users' brain signals is very weak and hard to synchronize accurately.

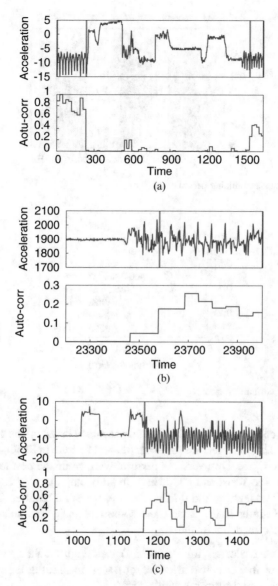

**Fig. 5.5** Finding periodic activities. To make periodic patterns clear, only part of traces in (**b**) and (**c**) is shown. (**a**) Acc-Wrist. (**b**) Acc-Chest. (**c**) Acc-PAMAP2

Figure 5.8a shows the synchronization accuracy for non-periodic traces using the best feature (i.e., the feature with the highest correlation coefficient), as described in Sect. 5.2. 100% activity traces and 57% P300 traces are synchronized within 15% normalized error.

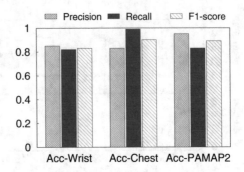

**Fig. 5.6**  Performance of finding periodic activities

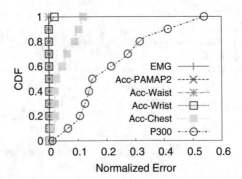

**Fig. 5.7**  CDF of normalized error

Figure 5.8b compares the synchronization accuracy based on the average correlation coefficient across all features or using the best feature. In EMG, Acc-PAMAP2, Acc-Wrist, Acc-Chest, both versions perform well, while the best feature performs much better in Acc-Waist and P300 EEG. In particular, the accuracy in Acc-Waist is 1 using the best feature, and 0 using the average of all features. This is because there are 591 features in the dataset and some features are so noisy that they can degrade the performance if included.

**Synchronization Accuracy for Periodic Traces**  Figure 5.9a shows the synchronization accuracy for periodic portions of the traces. In all cases, the synchronization accuracy is high, and the error is within 15%.

**Need of Separating Periodic and Non-periodic Traces**  So far we show our algorithms can synchronize periodic and non-periodic traces. To further demonstrate the need of using different synchronization algorithms for periodic and non-periodic traces, we apply the synchronization algorithm for non-periodic traces to a trace that contains both periodic and non-periodic data. As shown in Fig. 5.9b, the synchronization accuracy of the mixed traces is much worse than the synchronization accuracy of either periodic or non-periodic traces. This is because periodic traces need to be converted to the uniform frequency before finding the right offset that

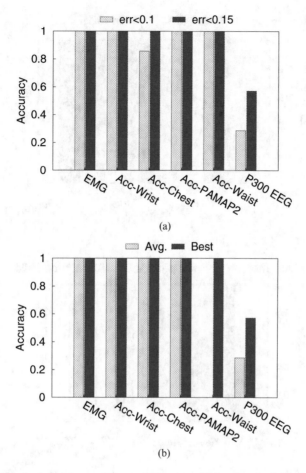

**Fig. 5.8** Synchronization accuracy of non-periodic traces. (**a**) Synchronization is accurate if error < *thresh*. (**b**) Avg. of features v.s. best feature

maximizes correlation coefficient. We also observe that periodic traces tend to have a higher synchronization accuracy because some non-periodic traces are flat and a slight shift may have little impact on the correlation coefficient.

## 5.4.2 Evaluation of Interpolation Accuracy

**Performance Metrics** We quantify the interpolation accuracy in terms of estimation error of the missing entries. We drop data from existing network matrices and compare our estimation with the ground truth. Normalized Mean Absolute Error (NMAE) is used to quantify the estimation error as follow:

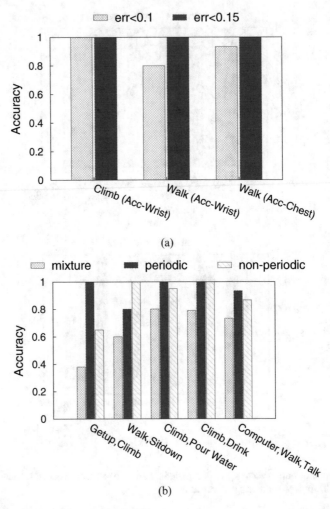

**Fig. 5.9** Synchronization accuracy of (**a**) periodic traces, and (**b**) a mixture of non-periodic and periodic traces where the first four sets of activities are from Acc-Wrist trace, and the last one is from Acc-Chest trace

$$NMAE = \frac{\sum_{i,j:M(i,j)=0} |X(i,j) - \hat{X}(i,j)|}{\sum_{i,j:M(i,j)=0} |X(i,j)|}, \qquad (5.3)$$

where $X$ and $\hat{X}$ are the original and estimated matrices, respectively. Only errors of the missing entries in the original matrices are compared. For each setting, we conduct 10 random runs, which drop a random set of data, and report an average of these 10 runs.

**Schemes Evaluated** Synchronization reduces the effective ranks of the matrices and can be applied to any interpolation algorithm. We apply it to the following three compressive sensing algorithms in our evaluation:

- SVD Base: It applies SVD to $X - X_{base}$, where $X_{base} = \overline{X} + X_{row}1^T + 1X_{col}^T$, 1 is a column vector consisting of all ones, and $X_{row}$ and $X_{col}$ are computed using a regularized least square according to [13]. This out-performs SVD applied directly to $X$, as shown in [13]. We observe similar results, so below we use SVD Base.
- SRMF: It uses Sparsity Regularized Matrix Factorization (SRMF) that leverages both low-rank and spatio-temporal characteristics [13].
- LENS: It uses LENS decomposition, which decomposes a matrix into a low-rank matrix, a sparse anomaly matrix, an error matrix, and a small noise matrix [3]. Main benefit of this algorithm is its robustness against anomalies.

**Varying Dropping Rates** We compare the interpolation accuracy in (i) the original trace, (ii) after a random shift, and (iii) after performing data-driven synchronization. We use the traces in Table 3.1. Figures 5.10, 5.11, and 5.12 show the interpolation errors of WiFi traffic trace for the three cases under SVD-Base, SRMF, and LENS, respectively. As the loss rate varies from 0.1 to 0.6, our synchronization reduces NMAE by 1% to 23%.

Figure 5.13a further plots the average amount of reduction in NMAE across three compressive algorithms under all traces. As we can see, the amount of reduction not only varies across traces, but also varies across interpolation algorithms. SVD-Base generally sees the highest benefit (around 7%–79%), followed by SRMF (around 8%–48%), and followed by LENS. The reduction in LENS is lowest because it can better cope with matrices of higher ranks. Nevertheless, even LENS sees 1%–30% reduction in NMAE after synchronization and 1%–44% increase in NMAE under a random shift, indicating the importance of synchronization. In addition, our interpolation supports both non periodic traces (the first 11 traces) and periodic traces (the last two traces). The reduction in interpolation error is comparable.

We also analyze the reduction in the interpolation error for the missing elements inferred during the first step. Compared with the reduction for all missing values shown in Fig. 5.13b, the error reduction in the first step is larger. The later steps experience larger error since we add extra missing values to transform misaligned portions into a regular matrix.

### 5.4.3 Evaluation of Segmentation

**Evaluation Methodology** We first demonstrate the effectiveness of our segmentation algorithm on non-periodic and periodic activities. Then we compare three algorithms: (i) our algorithm with synchronization, (ii) our algorithm without synchronization, and (iii) the existing algorithms. There are many variants of

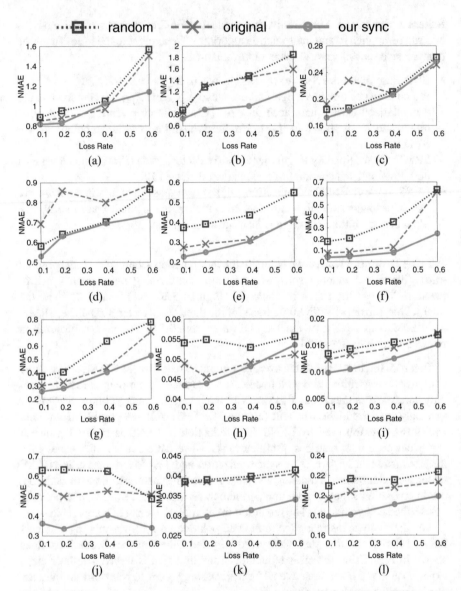

**Fig. 5.10** SVD-base: interpolation error under varying data loss rates after applying 20% random shift to each row and after our synchronization algorithm. (**a**) 3G. (**b**) WiFi. (**c**) Abilene. (**d**) GÉANT. (**e**) FourSquare Check-ins. (**f**) Multi-channel CSI. (**g**) Cister RSSI. (**h**) CU RSSI. (**i**) UMich RSS. (**j**) UCSB Meshnet. (**k**) 1-channel CSI. (**l**) Acc-Wrist:Walk

existing segmentation algorithms. We implement one variant that is based on first performing activity recognition and then look for the boundary between two different activities/events. Our implementation uses linear regression to perform activity recognition.

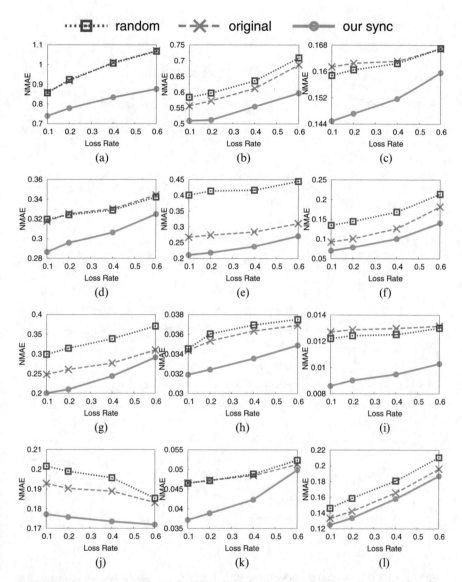

**Fig. 5.11** SRMF: Interpolation performance under varying data loss rates after applying 20% random shift to each row and after our synchronization algorithm. (**a**) 3G. (**b**) WiFi. (**c**) Abilene. (**d**) GÉANT. (**e**) FourSquare Check-ins. (**f**) Multi-channel CSI. (**g**) Cister RSSI. (**h**) CU RSSI. (**i**) UMich RSS. (**j**) UCSB Meshnet. (**k**) 1-channel CSI. (**l**) Acc-Wrist:Walk

We quantify the segmentation accuracy using precision and recall, where precision is the fraction of segmentation boundaries we detected are within 25% from the actual segmentation boundaries and recall is the fraction of real segmentation

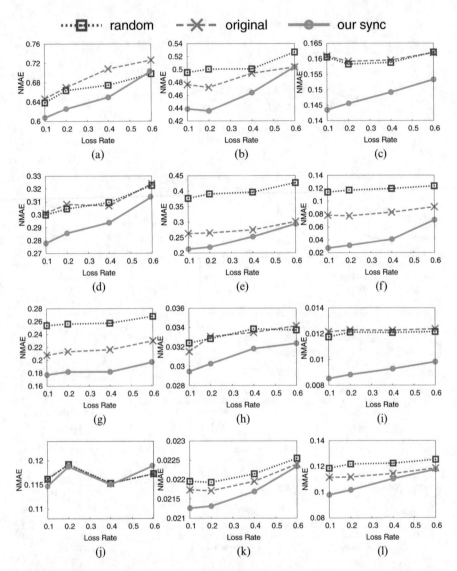

**Fig. 5.12** LENS: Interpolation performance under varying data loss rates after applying 20% random shift to each row and after our synchronization algorithm. (**a**) 3G. (**b**) WiFi. (**c**) Abilene. (**d**) GÉANT. (**e**) FourSquare Check-ins. (**f**) Multi-channel CSI. (**g**) Cister RSSI. (**h**) CU RSSI. (**i**) UMich RSS. (**j**) UCSB Meshnet. (**k**) 1-channel CSI. (**l**) Acc-Wrist:Walk

boundaries that are within 25% away from our detected segmentation boundaries. We further summarize these two metrics using *F1-score*, as defined earlier.

**Segmentation Accuracy for Non-periodic Traces** Figure 5.14a shows the segmentation results of non-periodic traces in terms of precision, recall, and F1-score.

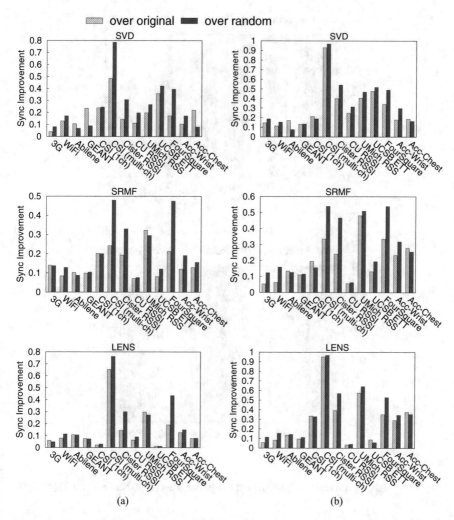

**Fig. 5.13** The interpolation error reduction of the missing values after synchronization against the original and the random-shift traces (10% loss rate) under (**a**) the entire trace and (**b**) the inner-most bounding box

The three metrics range between 0.75 and 1 for activity traces, and around 0.65 for P300 brain waves. P300 trace has lower precision and recall because EEG readings are subject to muscle movement and skin condition. The noisy data results in weak correlation coefficient so it is harder to find segments. Acc-Wrist trace has high recall but lower precision because some activities consist of several smooth portions with abrupt transitions (e.g., the non-periodic activity shown in Fig. 5.5a). The transitions result in high correlation coefficient and cause false positive.

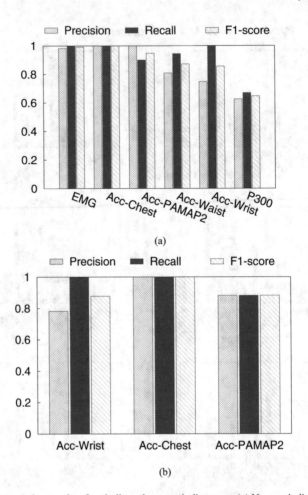

**Fig. 5.14** Segmentation results of periodic and non-periodic traces. (**a**) Non-periodic. (**b**) Periodic

**Segmentation Accuracy for Periodic Traces** Figure 5.14b shows the segmentation results of periodic traces. The F1-score is 1 for Acc-Chest, 0.88 for both Acc-PAMAP2 and Acc-Wrist. The segmentation accuracy is higher for periodic traces than non-periodic traces, because the latter may have flat portions and see similar correlation coefficient in the flat portions, thereby limiting the accuracy.

**Benefit of Separating Periodic and Non-periodic Traces** Figure 5.15 compares the segmentation accuracy of mixture traces versus either periodic or non-periodic traces. It is evident that mixture traces experience considerably lower accuracy. This confirms the importance of separating periodic and non-periodic traces before performing segmentation.

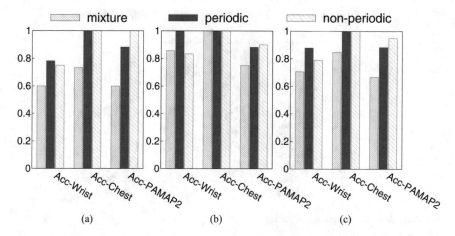

**Fig. 5.15** Segmentation results of the mixture traces. (**a**) Precision. (**b**) Recall. (**c**) F1-score

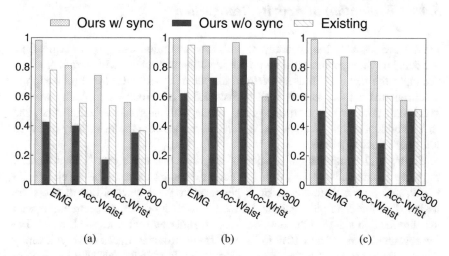

**Fig. 5.16** Segmentation results of different approaches. (**a**) Precision. (**b**) Recall. (**c**) F1-score

**Comparing with Other Segmentation Schemes** Figure 5.16 compares three algorithms: (i) our algorithm as presented in Sect. 5.3.2, (ii) our algorithm without synchronization, and (iii) the existing approach that first performs activity recognition and then performs segmentation. For all three traces, our approach achieves highest F1-score (a summary of precision and recall), and also highest precision and recall in most traces. The performance of (iii) is limited by the recognition accuracy. Recognition error may result in a false positive in segmentation, so it has high recall but low precision. (ii) yields the lowest precision because the the variation of correlation coefficients in the unsynchronized portions causes lots of false positives. This result shows that using multiple time series is useful for segmentation only when they are synchronized well.

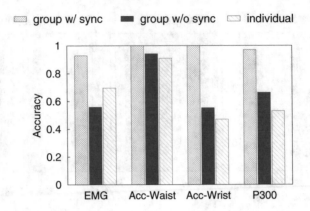

**Fig. 5.17** Activity recognition accuracy

## 5.4.4  Evaluation of Activity Recognition

Finally, we evaluate the accuracy of activity recognition using linear regression. We compare three schemes: (i) group-based voting with our synchronization, which leverages the fact that all users perform the same activity after synchronization to determine the activity as described in Sect. 5.3.3, (ii) group-based voting without synchronization, which performs group voting without synchronization first, and (iii) individual activity recognition which determines the type of activity based on one user's trace at a time.

Figure 5.17 compares the fraction of activities that are correctly identified by the three schemes. Note that voting is across different users in the accelerometer traces, and across different sessions per user in P300 since different users are assigned different target images. As we can see, group-based recognition with synchronization significantly out-performs the other two schemes in all cases. The improvement ranges from 10% to 113%. Unsynchronized traces can significantly degrade the accuracy of voting, and even under-perform individual activity recognition, as shown in EMG traces.

## 5.4.5  Summary of Evaluation Results

We summarize the evaluation results and observations as follows: (i) The synchronization error is 15% or less for 57–100% non-periodic traces and 100% for periodic traces. (ii) Our synchronization algorithm reduces the interpolation error by 1–79% for SVD, SRMF, and LENS. (iii) Separating period and non-period activities benefits synchronization accuracy and segmentation F1 scores by 18–163% and 12–42%, respectively. (iv) When there are multiple features, the synchronization based on the best feature (the one with highest correlation coefficient) out-performs that based on the average of all features by 0–99%. While for segmentation, using all features perform better because different activities may need different features to work well. (v) The error of the segmentation method enabled by our synchronization is 25% or less for 65–100% traces which is 12–61% better than the existing segmentation scheme. (vi) Our synchronization algorithm improves the activity recognition accuracy by 10–123%.

## 5.5  Existing Sychronization Approaches

A typical solution to achieve a network-wide synchronization is to align the clocks of distributed nodes [8, 12]. Another class of techniques is through monitoring packet reception [12] or protocol-specific features, such as a preamble. In addition, Lukac [7] proposes synchronization by utilizing background noise in seismic sensing systems. [1] proposes to detect events in data collected from ambient sensors or wearables, and then use these events for synchronization. Our work is the first that uses synchronization to reduce matrix ranks and improves compressive sensing. Moreover, we show it is important to distinguish periodic and non-periodic data and employ different synchronization algorithms accordingly.

## 5.6  Conclusion

In this chapter, we use extensive evaluation of real-world data to show data-driven synchronization plays a significant role in the data analytics. It can reduce matrix rank and improve the effectiveness of compressive sensing. To maximize its effectiveness, we develop (i) a simple yet effective synchronization algorithm that supports both periodic and non-periodic data, (ii) a new interpolation scheme that exploits improved synchronization to enhance accuracy, (iii) group-based segmentation and activity recognition algorithms that take advantage of synchronized activities across users. Our evaluation shows that synchronization reduces interpolation error by up to 79%, improves segmentation accuracy by 12–61%, and improves activity recognition accuracy by 10–123%.

## References

1. D. Bannach, O. Amft, P. Lukowicz,  Automatic event-based synchronization of multimodal data streams from wearable and ambient sensors,  in *Proc. of EuroSSC* (Springer, New York, 2009)
2. L. Bao, S.S. Intille,  Activity recognition from user-annotated acceleration data, in *Proc. of Pervasive*, 2004
3. Y.-C. Chen, L. Qiu, Y. Zhang, G. Xue, Z. Hu,  Robust network compressive sensing,  in *Proc. 20th Annu. Int. Conf. Mob. Comput. Netw. - MobiCom '14*, New York, New York, USA, September 2014 (ACM Press, New York, 2014), pp. 545–556

4. D. Garcia, Robust smoothing of gridded data in one and higher dimensions with missing values. Comput. Stat. Data Anal. **54**(4), 1167–1178 (2010)
5. Gets the extrema points from a surface (2022). http://www.mathworks.com/matlabcentral/fileexchange/12275-extrema-m--extrema2-m
6. T. Huynh, U. Blanke, B. Schiele, Scalable recognition of daily activities with wearable sensors, in *LoCA*, 2007
7. M. Lukac, P. Davis, R. Clayton, D. Estrin, Recovering temporal integrity with data driven time synchronization, in *Proc. of IPSN*, 2009
8. P. MacWilliams, B. Prasad, M. Khare, D. Sampath, Source synchronous interface between master and slave using a deskew latch, 2001. US Patent 6,209,072
9. G. Okeyo, L. Chen, H. Wang, R. Sterritt, Dynamic sensor data segmentation for real-time knowledge-driven activity recognition, in *Pervasive and Mobile Computing*, 2014
10. A. Rai, K.K. Chintalapudi, V.N. Padmanabhan, R. Sen, Zee: zero-effort crowdsourcing for indoor localization, in *MobiCom*, 2012
11. D. Roggen, K. Forster, A. Calatroni, The adARC pattern analysis architecture for adaptive human activity recognition systems, in *J Ambient Intell Human Computing*, 2013
12. B. Sundararaman, U. Buy, A.D. Kshemkalyani, Clock synchronization for wireless sensor networks: a survey. Ad Hoc Networks (Elsevier) **3**, 281–323 (2005)
13. Y. Zhang, M. Roughan, W. Willinger, L. Qiu, Spatio-temporal compressive sensing and internet traffic matrices. ACM SIGCOMM Comput. Commun. Rev. **39**(4), 267 (2009)

# Chapter 6
# Conclusion and Future Research Direction

In this monograph we focus on applying compressive sensing for network analytics and building systems that leverage the advantage of compressive sensing, such as anomaly detection, missing value interpolation, and future data prediction. We make important observations that real-world datasets often are not low-rank due to (i) the presence of measurement errors, noise, and anomalies, (ii) lack of synchronization in time domain and frequency domain. To our knowledge, this is the *first* work that shows the impact of noise, anomalies, and synchronization in the matrix ranks, which has profound implication for compressive sensing.

In Chap. 2, we develop a systematic method to automatically detect anomalies in a cellular network using the customer care call data. Our approach scales to a large number of features in the data and is robust to noise. Using evaluation based on the call records collected from a large cellular provider in US, we show that our method can achieve 68% recall and 86% accuracy, much better than the existing schemes.

In Chap. 3, we demonstrate the impact of noise, errors, anomalies, and lack of synchronization against the matrix ranks using a wide range of real-world data including network traces from 3G, WiFi, mesh, sensor networks, and the Internet, as well as activity traces from wearable devices. Our analysis show that these factors can make network matrices have a much higher rank. Violation of low rank assumption significantly reduces the effectiveness of existing compressive sensing approaches.

To address the problem of noise, errors, anomalies in the real-world data, in Chap. 4, we present *LENS decomposition* which decomposes a network matrix into a low-rank matrix, a sparse anomaly matrix, an error matrix, and a dense but small noise matrix. LENS has the following nice properties: (i) it is general: it can effectively support matrices with or without anomalies, and having low-rank or not, (ii) its parameters are self tuned so that it can adapt to different types of data, (iii) it is accurate by incorporating domain knowledge, such as temporal locality, spatial locality, and initial estimate, (iv) it is versatile and can support many applications including missing value interpolation, prediction, and

© The Author(s), under exclusive license to Springer Nature Switzerland AG 2022
G. Xue et al., *Robust Network Compressive Sensing*, SpringerBriefs in Computer Science, https://doi.org/10.1007/978-3-031-16829-1_6

anomaly detection. Our evaluation shows that it can effectively perform missing value interpolation, prediction, and anomaly detection and out-perform state-of-the-art approaches.

To address the problem of the lack of synchronization, in Chap. 5, we present a data-driven synchronization approach to explicitly remove misalignment while accounting for the time domain and frequency domain heterogeneity of the real-world data. Data-driven synchronization reduces the matrix ranks by up to 65% and improve the interpolation performance of all compressive sensing methods. Moreover, we show that data-driven synchronization can enable group-based segmentation and activity recognition which significantly out-performs existing schemes which applied on an individual user at a time.

# Appendix A
# Alternating Direction Method

## A.1 Algorithm

Each iteration of the Alternating Direction Method involves the following steps:

1. Find $X$ to minimize the augmented Lagrangian function $\mathcal{L}(X, \{X_k\}, Y, Y_0, Z, W, M, \{M_k\}, N, \mu)$ with other variables fixed. Removing the fixed terms, the objective is:

$$\text{minimize: } \alpha\|X\|_* + \frac{\mu}{2} \sum_{k=0}^{K} \|X_k + M_k/\mu - X\|_F^2.$$

Let $J = \frac{1}{K+1} \sum_{k=0}^{K}(X_k + M_k/\mu)$, and $t = \frac{\alpha/\mu}{K+1}$. We can simplify the objective to the following:

$$\text{minimize: } t\|X\|_* + 1/2\|X - J\|_F^2.$$

According to matrix completion literature, this is a standard nuclear norm minimization problem and can be solved by applying soft thresholding on the singular values of $J$. Specifically, we have:

$$X = \mathsf{SVSoftThresh}(J, t).$$

For a given $J$ and $t$, let $J = USV^T$ be the singular value decomposition of $J$. We have: $\mathsf{SVSoftThresh}(J, t) \overset{\triangle}{=} U\,\mathsf{SoftThresh}(S, t)V^T$, where

$$\mathsf{SoftThresh}(S[i, j], t) \overset{\triangle}{=} \mathrm{sign}(S[i, j]) \max(0, |S[i, j]| - t).$$

© The Author(s), under exclusive license to Springer Nature Switzerland AG 2022
G. Xue et al., *Robust Network Compressive Sensing*, SpringerBriefs in Computer Science, https://doi.org/10.1007/978-3-031-16829-1

2. Find $X_k$ to minimize $\mathcal{L}(X, \{X_k\}, Y, Y_0, Z, W, M, \{M_k\}, N, \mu)$ with other variables fixed ($k = 1, 2, \ldots, K$). This gives:

$$\text{minimize: } \frac{\gamma}{2\sigma} \| P_k X_k Q_k^T - R_k \|_F^2 + \frac{\mu}{2} \| X_k + M_k/\mu - X \|_F^2.$$

This is a least square problem with respect to $X_k$. The optimal solution can be obtained by forcing the gradient of the objective to be zero. That is,

$$\frac{\gamma}{\sigma} P_k^T (P_k X_k Q_k^T - R_k) Q_k^T + \mu(X_k + M_k/\mu - X) = 0. \tag{A.1}$$

Let $J = X - M_k/\mu$, and $R = P_k^T R_k Q_k + \frac{\mu\sigma}{\gamma} J$. Eq. (A.1) simplifies to

$$P_k^T P_k X_k Q_k^T Q_k + \frac{\mu\sigma}{\gamma} X_k = R. \tag{A.2}$$

Perform eigendecomposition on $P_k^T P_k$ and $Q_k^T Q_k$ and let $U S U^T = P_k^T P_k$; $V T V^T = Q_k^T Q_k$, where $U$ and $V$ are orthogonal matrices, $S$ and $T$ are diagonal matrices. We have: $S(U^T X_k V)T + \frac{\mu\sigma}{\gamma} (U^T X_k V) = U^T R V$. Through a change of variable, let $H = U^T X_k V$, Eq. (A.2) becomes:

$$SHT + \frac{\mu\sigma}{\gamma} H = U^T R V. \tag{A.3}$$

Let $\mathbf{s} = diag(S)$, $\mathbf{t} = diag(T)$ be the diagonal vector of $S$ and $T$, respectively. Eq. (A.3) is equivalent to $(\mathbf{st}^T + \frac{\mu\sigma}{\gamma}). * H = U^T R V$. Since $U$ and $V$ are orthogonal matrices, we can easily find $X_k$ from $H$ as $X_k = U H V^T$. So we have: $H = (U^T R V)./(\mathbf{st}^T + \frac{\mu\sigma}{\gamma})$, where $./$ is an operator for element-wise division. Thus,

$$X_k = U H V^T = U \left( (U^T R V)./(\mathbf{st}^T + \frac{\mu\sigma}{\gamma}) \right) V^T.$$

3. Find $X_0$ to minimize $\mathcal{L}(X, \{X_k\}, Y, Y_0, Z, W, M, \{M_k\}, N, \mu)$ with other variables fixed ($k = 1, 2, \ldots, K$). This gives:

$$\begin{aligned}
\text{minimize: } & \langle M, D - AX_0 - BY_0 - CZ - W \rangle \\
& + \langle M_0, X_0 - X \rangle \\
& + \mu/2 \| D - AX_0 - BY_0 - CZ - W \|_F^2 \\
& + \mu/2 \| X_0 - X \|_F^2
\end{aligned}$$

That is,

$$\text{minimize: } \|X_0 - X + M_0/\mu\|_F^2$$
$$+\|D - BY_0 - CZ - W + M/\mu - AX_0\|_F^2$$

Let $J_0 = X - M_0/\mu$ and $J = D - BY_0 - CZ - W + M/\mu$. It becomes:

$$\text{minimize: } \|X_0 - J_0\|_F^2 + \|AX_0 - J\|_F^2$$

Letting the gradient be zero leads to: $X_0 - J_0 + A^T(AX_0 - J) = 0$. Therefore, $X_0 = \text{inv}(A^T A + I)(A^T J + J_0)$.

4. Find $Y$ to minimize $\mathcal{L}(X, \{X_k\}, Y, Y_0, Z, W, M, \{M_k\}, N, \mu)$ with other variables fixed. This gives:

$$\text{minimize: } \beta\|Y\|_1 + \langle N, Y_0 - Y \rangle + \mu/2\|Y_0 - Y\|_F^2$$

That is:

$$\text{minimize: } \beta/\mu\|Y\|_1 + 1/2\|Y_0 + N/\mu - Y\|_F^2.$$

Let $J = Y_0 + N/\mu$. $t = \beta/\mu$. It becomes: $t\|Y\|_1 + 1/2\|J - Y\|_F^2$. This can be easily solved as $Y = \text{SoftThresh}(J, t)$. To see why, the problem can be solved for each element of $Y$ separately. So we just need to find $Y[i, j]$ that minimizes: $t|Y[i, j]| + 1/2(J[i, j] - Y[i, j])^2$.

5. Find $Y_0$ to minimize $\mathcal{L}(X, \{X_k\}, Y, Y_0, Z, W, M, \{M_k\}, N, \mu)$ with other variables fixed. This gives:

$$\text{minimize: } \langle M, D - AX_0 - BY_0 - CZ - W \rangle$$
$$+\langle N, Y_0 - Y \rangle$$
$$+\mu/2\|D - AX_0 - BY_0 - CZ - W\|_F^2$$
$$+\mu/2\|Y_0 - Y\|_F^2$$

Let $J_0 = Y - N/\mu$, $J = D - AX_0 - CZ - W + M/\mu$. It becomes minimize: $\|Y_0 - J_0\|_F^2 + \|BY_0 - J\|_F^2$. Letting the gradient $= 0$, we obtain: $Y_0 - J_0 + B^T(BY_0 - J) = 0$. So $Y_0 = \text{inv}(B^T B + I)(B^T J + J_0)$.

6. Find $Z$ to minimize $\mathcal{L}(X, \{X_k\}, Y, Y_0, Z, W, M, \{M_k\}, N, \mu)$ with other variables fixed. This gives:

$$\text{minimize: } \frac{1}{2\sigma}\|Z\|_F^2 + \langle M, D - AX_0 - BY_0 - CZ - W \rangle$$
$$+\frac{\mu}{2}\|D - AX_0 - BY_0 - CZ - W\|_F^2.$$

Let $J = D - AX_0 - BY_0 - W + M/\mu$, it becomes:

$$\frac{1}{2\mu\sigma}\|Z\|_F^2 + \frac{1}{2}\|CZ - J\|_F^2.$$

Letting the gradient = 0 yields: $Z = \text{inv}(\frac{1}{\mu\sigma}I + C^T C)(C^T J)$.

7. Find $W$ to minimize $\mathcal{L}(X, \{X_k\}, Y, Y_0, Z, W, M, \{M_k\}, N, \mu)$ with other variables fixed. This gives:

$$\text{minimize:} \quad \langle M, D - AX_0 - BY_0 - CZ - W\rangle$$
$$+\mu/2\|D - AX_0 - BY_0 - CZ - W\|_F^2.$$

That is:

$$\text{minimize:} \quad \|D - AX_0 - BY_0 - CZ - W + M/\mu\|_F^2$$

So $W = E. * (D - AX_0 - BY_0 - CZ + M/\mu)$ (recall that $W = E. * W$).

8. Update estimate for $\sigma_D$ as follows. Let $J = D - AX_0 - BY_0 - W$. We then compute $\sigma_D$ as the standard deviation of $J[E = 0]$ and update $\sigma = \theta\sigma_D$. In our implementation, we fix $\theta = 10$.

9. Update estimates for the Lagrangian multipliers $M$, $M_k$ and $N$ according to: $M = M + \mu \cdot (D - AX_0 - BY_0 - CZ - W)$, $M_k = M_k + \mu \cdot (X_k - X)$ $(k = 0, \cdots, K)$, $N = N + \mu \cdot (Y_0 - Y)$.

10. Update $\mu = \mu \cdot \rho$. In our implementation, initially $\mu = 1.01$ and $\rho = 1.01$. Every 100 iterations, we multiply $\rho$ by 1.05.

Printed in the United States
by Baker & Taylor Publisher Services